A Guide to Fleet Management and Company Cars

A Guide to Fleet Management and Company Cars

Les Cheesman
Anthony Minns

Croner Publications Ltd
Croner House
London Road
Kingston upon Thames
Surrey KT2 6SR
Telephone: 081-547 3333

Copyright © 1993 L Cheesman and A Minns
First published 1993

Published by
Croner Publications Ltd,
Croner House,
London Road,
Kingston upon Thames
Surrey KT2 6SR
Telephone: 081-547 3333

The right of L Cheesman and A Minns to be identified as authors of this work has been asserted by them in accordance with the Copyright, Designs and Patents Act 1988.

All rights reserved.
No part of this publication may be reproduced, stored in a retrieval system, or transmitted in any form or by any means, electronic, mechanical, photocopying, recording or otherwise, without the prior permission of Croner Publications Ltd.

While every care has been taken in the writing and editing of this book, readers should be aware that only Acts of Parliament and Statutory Instruments have the force of law, and that only the courts can authoritatively interpret the law.

British Library Cataloguing in Publication Data
A CIP Catalogue Record for this book is available from the British Library.

ISBN 1 85524 119 6

Printed in Great Britain by
Whitstable Litho Printers Ltd, Whitstable, Kent

Contents

		Page
Chapter 1	Introduction	1
Chapter 2	Company Car Policy	7
Chapter 3	Car Fleet Allocation	15
Chapter 4	The Car Fleet Budget	21
Chapter 5	Acquisition of Cars	27
Chapter 6	Car Cost Control	43
Chapter 7	Disposal of Cars	51
Chapter 8	Taxation of Company Cars	57
Chapter 9	Buying Out the Company Car	85
Chapter 10	The Future for the Company Car	91
Appendix	Specimen Employee Booklet for the ABC Company Car Scheme	97
Index		105

THE AUTHORS

Les Cheesman

Les Cheesman joined Henley Management College full time as Deputy Director, Centre for Automotive Management, in June 1990, having spent over 20 years in transport-related industries.

He held senior posts in a number of major corporations which included the NFC (Tankfreight) as Marketing Manager, Freightliners (British Rail) as General Manager (Road), Wincanton Transport as Director and Sea Containers Group (Trafpak Services) as Managing Director. Now a visiting faculty Member of Henley Management College, his involvement also encompasses additional work from Henley Distance Learning Ltd and other consultancies.

His experience covers most aspects of automotive management from truck rental, leasing and contract hire of cars, vans and trucks to intermodal movement by rail, road or ship and freight forwarding of ISO containers, both as a provider and user.

Les has lectured on aspects of transport at a number of conferences and contributed to published books and articles on Fleet Management, Energy Saving and Tractors for haulage application.

Anthony Minns

Tony Minns is currently Managing Director of MM & K Limited, a company which specialises in tax and management remuneration issues. MM & K Limited was acquired by its directors and executives from three large corporate shareholders, Merchant Bank — Morgan Grenfell, Insurance Broker — Willis Faber and Management Consultant — PA Consulting.

After being called to the Bar in 1969, Tony Minns has worked as a UK legal advisor to food manufacturer, CPC International; for ten years at Morgan Grenfell and as a director of Bank of America International before acquiring his stake as former shareholder in the management buy-out of MM & K.

Tony Minns is one of the two authors of *Kluwer's Effective Remuneration*, has written a number of other tax reference works, sits on the editorial panel of a number of Croner publications and lectures on a variety of tax-based topics.

Chapter 1
INTRODUCTION

FLEET MANAGEMENT

Terminology within the car fleet industry is confused and few standards have yet been set. Whilst many would consider "fleet management" to focus only on the companies that provide these services to third parties, there is an increasing band of managers who manage fleets for their own companies.

Facts are hard to come by and those who claim to know the facts are found to be not necessarily in tune with their colleagues.

Conferences held on the subject of company car fleet management fail to agree on definitions, market size or direction but, in general terms, it would seem that everyone agrees there were about 4.5 million "business" cars on the road in 1991, of which over half were driven by the self-employed. In many respects, these self-employed cars are no less a company car as they are cars that are operated "on the business" for small businesses in their own right.

COMPANY CARS

A company car can be defined as a car provided by an employer that is available for private use by an employee, or one to which the "scale charges" apply, that are set by the Inland Revenue.

The Inland Revenue estimates that in 1991 there were 1.9 million company cars operating and the best estimate of the nature and structure of company car fleets in the UK against a series of price bands was given as follows:

Price Band £	Cars in Use (Thousands)
5,000 – 6,499	10
6,500 – 7,999	50
8,000 – 9,999	220
10,000 – 12,499	800
12,500 – 15,499	400
15,500 – 19,499	310
19,500 – 24,999	60
25,000 – 30,999	30
31,000 – 38,999	20
39,000 – 49,000	5
	1905

The number of company cars has grown markedly over the last few years from less than 500,000 in the mid-1970s, as British companies, more than most European counterparts, embraced the concept of the company car as part of the total remuneration package for a wide range of employees.

Whether an essential tool for the trade or a necessary inducement to encourage the best quality staff to join and subsequently stay with the company, "the company car" has caused the Government some concern. It has fought to adopt a policy of tax neutrality to ensure that those employees who do not have a company car are not unduly disadvantaged by those who do and that those who have a company car pay for it, in terms of a benefit-in-kind, in a manner that is equitable with the true benefit they obtain.

THE INDUSTRY

In 1988, there were estimated to be 18.4 million cars in Britain and it is thought that by the year 2000 there will be up to 24 million. In 1989, some 2.2 million cars were sold, in 1990, some 1.9 million and in 1991, 1.6 million. With the general recession taking a firm grip in 1992, the figure is unlikely to exceed the 1991 figure.

Around 63% of all new cars sold in the UK are operated and run in some way "on the business". It is also considered that some 50% of all new cars sold are purchased for the conventional "company car" fleet use.

Therefore business cars in general and company cars in particular are a significant part of the British economy.

It is estimated that company cars produce an income tax yield of about £1.4 billion pa from the car benefit scales alone. Therefore any changes to the manner in which company cars are dealt with can have a significant impact on the British economy.

There are *three key issues* that need to be considered.

1. *The Government* generates considerable revenue from the car benefit scale charges, as well as from fuel tax and VAT.
2. *The car manufacturing industry* has to ensure that cars meet the specification likely to attract new customers whilst being produced within the most advantageous parameters determined by legislation such as price, engine capacity or fuel type. Changes in legislation can upset their long-range planning.
3. *The company car driver* will use a wide range of parameters on which to make a choice. Price may be influenced by the employer; performance may be a personal influence; size and comfort along with specification and style will also influence choice.

Whilst the employer will be looking possibly for best value in terms of whole-life cost and economic operating (and perhaps influenced by "green" issues), the employee will probably view the company car as simply part of the pay package.

FLEET MANAGERS

It has been estimated that there are over 20,000 fleet managers running fleets of over 25 vehicles and many of these managers, apart from the very large operators, run the fleet part time with little or no training or knowledge of the industry other than that passed on by their predecessor.

Company cars are always a very emotional topic. There is much talk of the economy of diesel and down-sizing to improve economy to the employer and the improved specification to satisfy status to the employee. However, much of the pressure placed on the fleet manager comes from the fact that the company car fleet drivers see the vehicles as a personal possession and as a direct manifestation of their lifestyle. A fleet manager therefore has the difficult task of balancing these two contrasting forces — economy for the company and maximum satisfaction for the employee.

Around August each year when, historically, many car fleets are changed to take advantage of the new registration letter prefix (yet another sign of status), the fleet manager can be the most popular person around. If, however, the management decision has already been taken to run the vehicles for another year then the fleet manager's task will be that much more difficult to do.

Although, in theory, the car fleet manager is the master of the company car policy, in practice he or she may have no budget or no money allocated, despite previous agreements. There may be a demand to cut the budget by at least 10% from last year's despite being well within this year's budget and immediate demands on the manager for services outside the agreed standards, without any chance to control costs. When it comes to disposal, often little prior planning with the fleet manager is involved, ie people leave ... cars stay!

THE FIGHT BACK

As the fleet manager's role is often perceived as a minor one in the company, positive steps to provide pro-active management information are required. By raising the profile of the fleet responsibility and taking every opportunity to raise issues of a constructive nature, the fleet manager can improve the value of the job and increase his or her managerial profile. It is important to develop communication skills with all the vehicle users and as these may often involve senior managers, all communication should be clear, concise, accurate, and well presented, both written or verbally delivered at management meetings and conferences.

Finally, the fleet manager should seek training to ensure that his or her management techniques are up to date and get involved with whichever local professional institute has particular relevance to his or her business, ie CIT, IOTA, IPM, IMI, FTA, etc.

A fleet manager will have many contacts with the industry and if these contacts are carefully nurtured and developed, considerable insight into the way the fleet industry operates and the benefits that can accrue to the company will be gained.

It is worth noting, however, that there are also associations between many of the suppliers of services and the industry itself and a small example of the links is given in the chart overleaf.

There is nothing suspicious or underhand about this association. It exists and although there are legal restrictions on the use of information held in computer databases, it is as well to be aware of these links which explain why some companies may have more knowledge of a car fleet operation than the manager might have thought possible.

A Guide to Fleet Management and Company Cars

Figure 1
Some of the Links Between Companies in the Motor Industry

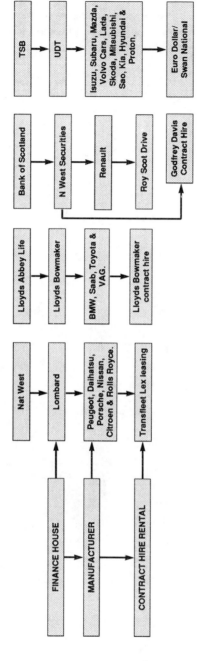

Chapter 2
COMPANY CAR POLICY

SETTING OUT THE POLICY

Generally, fleet managers do not start from a clearly defined base position and few have the luxury of starting up a company car policy from scratch. So, the first things to establish are the ground rules. These can be most easily identified by dividing them into three separate subheadings:

(a) where is the company car policy?
(b) where would the company like that policy to be?
(c) how will the company achieve its goal?

This may seem a simple task but it is by far the most important job to be done by a newly appointed fleet manager. Until there is a clear, detailed position statement of where the company is now, progress cannot be made. Many decisions within the company, prior to this review, may not have been clearly (or well) made, resulting in some reluctance on the part of senior managers to make their position clear.

A checklist can be helpful to provide milestones for measuring progress. The following may prove useful.

WHERE IS THE COMPANY CAR POLICY?

There are a number of key questions that need to be addressed which may require decision-making at the highest level. These include:

1. *Should the company provide cars?*
   If so, at what level of employment within the company should individuals be provided with a car?

2. *What is the basis for setting that level?*
   - salary?
   - job related?
   - need?
3. *What type or level of cars could be provided?*
4. *How should the car fleet be considered?*
   - should it be based on whole-life costs?
   - on the monthly charge from a contract hire company?
   - on the list price from the vehicle manufacturer or perhaps on an alternative cash option?
5. *What should the car fleet structure be?*
   - uniform fleet throughout the whole corporation?
   - focused fleet meeting the separate needs of both the head office staff and those travelling long distances in the field?
6. *Should drivers choose their own vehicles and how flexible should the company be?*
   - should there be limitations on manufacturers?
   - what has the policy been so far?
7. *Does the company have a fleet management strategy?*
   - is the fleet managed with own in-house resources?
   - does the company run the fleet or should it subcontract everything? Should the company subcontract part or manage some of the fleet in-house?
   - should there be an amalgamation with other departments within the company?
   - if the company has a mixture of cars and trucks, how should the workshop costs be allocated?
8. *What is the fleet management strategy overall?*
   - does the company have a long-term desire to provide vehicles?
   - does the company have the funds necessary to acquire them?
   - does the company have the management expertise necessary to control them?
   - having obtained the vehicles are they used effectively? Is their use controlled or are there serious cost implications which cannot be accounted for?
   - are they disposed of effectively when the company has no further use for them?
9. *What is current fleet funding strategy?*
   - is it part of the total financial strategy of the company; an integral part of the business?
   - does the company focus on the minimum cost for the maximum

use of vehicles?
- is the company concerned whether the funding is on or off the balance sheet?
- is there a clear policy on fleet funding or does it rather depend on the circumstances in which the rest of the company funds itself?
10. *What is the strategy regarding externally purchased services?*
    - if the company buys in fleet management services, what are the criteria for doing so?
    - does the company need to use pool cars or should it have a daily rental contract for the occasional use of extra vehicles?
11. *Is it important where the company sources its finance and are its lines of credit sometimes restricted?*
12. *When the company acquires a new company, does it have any strategic plan, eg keep the two fleets separate or amalgamate them immediately?*

With the capital cost of the average company car from £10,000 to over £25,000, and the operating costs for each vehicle varying from between £4,000 to £12,000 pa, the provision of company cars is a significant drain on the company's resources.

In practice, it is found that lack of proper control can mean that some vehicles may cost up to 20% more to run than other (identical) vehicles in the fleet, representing significant waste. The waste on a vehicle fleet of (say) 200 vehicles could mean spending over £100,000 pa more than necessary with an inevitable adverse impact on annual profits.

WHERE WOULD THE COMPANY LIKE THAT POLICY TO BE?

It would be sensible to try and set five clear objectives to meet company needs. These objectives could be taken from the following list:

- the need to attract the best calibre staff who have an expectation of being provided with a company car
- to provide essential users with cars suitable for the work they do
- to limit severely the number of company cars operating to those who really need them
- to provide cars that reflect the mileage run on company business in order to minimise fatigue of the driver
- to reflect the status of individuals within the company

- to enable those members of staff who do not want company cars to be offered an equal cash alternative.

The list can go way beyond these issues and should reflect those most significant to each company. Once established, and agreed by senior management, the process by which these objectives are to be met can be more clearly identified.

To determine where a company wants to be, it is useful to consider the implications of company car policy under clearly defined headings. This can help to focus attention on the issues that are most significant at the time and can further help a review of the situation in subsequent years, based on the criteria that were operating at the time and the changes that have occurred since then.

A way of doing this is to choose four broad "subject" headings that cover the range of issues most likely to affect a business. A clear choice is to consider the political, economic, social and technological changes that could influence where a company would like the car policy to be.

Political Issues

Both Government and internal politics could be reviewed in terms of the manner in which the Government of the day is treating the provision of company cars and the tax implications both to the individual and to the company. Where changes are being considered, as at present, clearly defined views could be taken based on the best expectation. The company policy could then be altered in line with the broad political spectrum. The internal politics of the company should also be considered in terms of the identified needs within the organisation.

Economic Issues

It is worth considering the future economics of the company in terms, perhaps, of access to new lines of credit and the need to use the funds currently employed in providing the cars for some other purpose within the company more directly related to profits.

Social Issues

This heading could include the needs of staff in terms of the recruitment policy for the future and reflect, perhaps, the changing nature of the organisation where more staff may be home based or work from regional centres. Additionally, the broader issues such as the environmental considerations that are becoming much more significant in companies' policy decisions and the need to ensure that the cars are efficient and generate the minimum of pollution which would result in choosing energy efficient engines with low emission, perhaps diesel for economy and long life and designed to be "recyclable".

Technology Issues

Apart from the specification of the cars, the increased sophistication of car fleet management through computers and in-board diagnostic equipment means that more preventative maintenance can be done and costs minimised. Also, with good technological packages on simple computers, good fleet management control means that rogue vehicles could be identified quickly should they occur and driver abuse identified and dealt with by the manager as appropriate.

Having established where company car policy is now and where it should be, all that is now required is the process by which it can be achieved.

HOW WILL THE COMPANY ACHIEVE ITS GOAL?

Reviews of management strategy regularly carried out by consultancy firms over a broad range of industries identify that the majority of strategic decisions are based on a hunch; very few companies use a structured approach. It would seem that companies do not compare their results against their major competitors and although many companies concede that their information base is suspect, do little to change it.

Fleet managers should ensure that they have sufficient knowledge about the products in the industry and should approach the industry directly and seek quotations for various schemes that would meet the company's key objectives.

Making a Start

At least three suppliers of any given "product" should be approached to obtain comparative results. Perhaps the most hopeful approach would be to provide chosen suppliers in the industry with details of the company's current position and its objectives so that they can match the company's need with their product. This may elicit at least some positive response.

The second approach would be for fleet managers to discuss with industry colleagues the methods which they use to achieve cost effective control of their fleet.

Alternatively, consultants in the industry may be able to advise on and provide an unbiased balanced view of a path or series of paths to follow.

It is important to get the strategic implications of the policy right and many companies have failed to achieve their desired objectives simply because the fundamental business decision was based on the wrong structure.

Fleet managers need to consider the following aspects.

Operational aspects

This would include the decision on control systems, policy on tyres, fuel and the contents of the fleet handbook.

Commercial aspects

This would include company cost implications and their effect on the core business of running the company fleet of vehicles and also the support available from any in-house systems that are relevant to the management of the fleet.

Tactical aspects

This would include such areas as who sets the policy or plan and when the plan is to be implemented and the resources that are to be made available and their timing.

Strategic aspects

Over all these considerations must lie the structure of the fleet, the funding and its management and whether the policy should be to have the funding on or off the balance sheet and the significance this may have to the rest of the operation of the company.

SOME OF THE BROADER ISSUES

Underlying many of the key issues will be the recurring problem — how should the company define the nature of the car to be provided. Is it work or "perk"?

It is important to establish precisely the use to which the company car will be put as this will have driver and company tax implications.

There are three typical car profiles which are as follows:

1. *The business traveller* who does, say, 25,000 business miles a year, perhaps 8,000 private miles, moves samples, uses his car as a mobile office, is home based and creates no major image impact for the company.
2. *The operational manager* who perhaps drives 3,000–20,000 business miles each year plus perhaps 12,000 private and has some equipment to be moved, is office based and remote from home. The car may have more image significance to the company as it is used to take clients out.
3. *The senior manager* who does perhaps nil or maybe 5,000 business miles each year plus perhaps 14,000 miles private use. They may travel to work by train, will often never move equipment and may create a major image impact for the company, needing to be seen in a new vehicle of high specification.

Unfortunately, the current Inland Revenue approach to company cars is to consider only the business use in terms of the taxable benefit-in-kind, whereas perhaps it should be the private use which is of more benefit to the individual and on which the individual would be more willing to pay tax.

Chapter 3
CAR FLEET ALLOCATION

This is a thorny issue at the best of times, and a subject which needs clear guidelines and a firm stance from all managers so that any exceptions that are allowed are included in the guidelines to satisfy both employee and employer.

A good way to start the process is to create a grading/seniority profile within the company (if none exists) that operates across the company (and perhaps its subsidiaries). This should include all personnel entitled to a car, based on a review that started on the understanding that no one was entitled to a car. Unless a "zero car" option is reviewed, this process is not complete and will lead to an untidy solution.

Having completed the first review, a list can be compiled which might look as follows:

Grade	Classification
1	Main Board Director
2	Senior Managers
3	Middle Management/Operations and Sales
4	Sales Managers
5	Operations & Sales Staff

Once the grades have been established (five should be enough) the selection of cars to fit the grades can be started.

Based on published information, or a company's own data, a "whole-life" cost can be calculated that takes into account initial purchase price, maintenance cost, insurance, fuel consumption and residual value.

It is generally considered unwise to focus on specific makes or models of vehicle. It is better to give a price band within which employees can choose whichever model they prefer but this can lead to higher cost for the company by models being chosen with poor residual values or high maintenance costs. It may well be better for the fleet manager to determine a choice of around four or five cars from within each price band where an attempt has been made to determine the best overall cost of ownership (see *Whole Life Costs* in Chapter 6).

Once this list has been compiled, the potential drivers can visit their local dealers and select the vehicle they prefer within the relevant price band. Then for companies using contract hire facilities, a search could be undertaken to see who would offer the best terms for that particular vehicle.

If contract hire with maintenance is a choice, it may well be possible to negotiate with the contract hire company a convenient garage to provide the basic servicing. Also, what *local* arrangements can be made to provide back-up in the event that servicing cannot be carried out at the weekend or the company car driver must have a vehicle at all times.

Car specifications change considerably over time and it is therefore difficult to give a definitive list of the type of cars which are likely to fall into these individual grades. The table opposite is an example of what could be included.

Specifying the Business Car

Car specification is vital if a company is to get the best value for money and the drivers get a vehicle they will appreciate and treat carefully.

There are three major areas to be considered:

(a) the status of cars within the organisation
(b) the vehicle selection and specification
(c) the advice the fleet manager can provide to encourage the potential drivers to make the right choice.

The Process of Allocation

This will depend on the nature of the company. It is possible that the parent company, controlling a group of subsidiaries, will dictate policy on salary and company car allocation that applies across the whole group.

Alternatively, it may be possible to consider, quite distinctly, those cars provided as a direct need to the business and those provided as part of a salary package where little or no company use is made of the vehicle, ie a "perk" car.

Vehicles can be allocated by grade, which is often used as a benchmark, or by income, which is probably best avoided as the allocation process is

Possible Car Grade Scheme
(based on car type)

Grade 1	Rover 820 Si 16v
	BMW 320i
	Granada Ghia 2.0i
	Volvo 940 SE Estate Auto
	Carlton 2.0 GLi Estate
	Rover 825 Turbo Diesel*
Grade 2	BMW 318i
	Granada 2.0i LX
	Volvo 740 GL Estate Man.
	Granada 2.5 DLX Turbo
Grade 3	Cavalier 2.0 GLi
	Rover 416 GLi
	Volvo 460 GLEi
	Sierra 2.0i GLS
	Peugeot 405 1.9 GRD Estate*
Grade 4	Rover 214 s
	Cavalier 1.8L
	Sapphire 1.8 LX
	Volvo 460 GL
	Cavalier 1.7 DL*
Grade 5	Astra Merit 1.2i
	Maestro 1.3
	Peugeot 309 1.3

Notes:
1. A car can be selected from a band above entitlement if the diesel option (*) is selected, as this creates cost savings for the company.
2. Start and finish point, ie Grade 5 to Grade 1, will be determined by the company's strategic policy. For example, if you start with a Rolls Royce in Grade 1, it gives you more scope in the other grades.

sufficiently emotional without additional conflict.

Finally, there is the so called "user-chooser" system where the individual has a choice within a price band. It would be advisable to restrict car choice, as certain models suffer more damage and are more prone to being stolen than other more "conventional" models. Ultimately, this decision will depend upon the ability of the company to determine the fundamental reason why they are providing company cars in the first place. Once that is known, the fleet manager can make recommendations to senior management of likely cause and effect.

Vehicle Selection and Specification

The fleet car needs to be specified within the range according to the needs of the driver. The items to be included are:

- number of doors
- the style of the body
- the engine capacity and type
- the gearbox
- colour.

The fleet manager would probably wish to give as wide a choice as possible to the potential driver but it must be remembered that every decision made at this stage may have a significant effect on the whole-life operating costs of the vehicle and particularly on its residual value. For example, it is said that the easiest car to sell second-hand is a red one.

Advice to Users

To provide advice that is not solicited by the potential driver is probably a dangerous pastime but, if possible, drivers should be encouraged to talk to the fleet manager well ahead of their decision to take particular models so that the fleet manager has the opportunity to influence the decision. Some of the items to be considered are:

1. What do users want from a car?
2. Can their company and private needs be analysed in some way?
3. Have they taken a road test and obtained details from the manufacturers?
4. Are they aware of the operating costs and their tax liabilities?
5. What image do they wish to promote for themselves and the company?

The important thing is that the influence the fleet manager can have on a sensible choice of vehicle at this stage and one that fits in with the company's budgetary requirements will be much easier if the individual has not made up his or her mind beforehand.

Chapter 4
THE CAR FLEET BUDGET

THE BUDGET PROCESS

For the fleet manager, the annual budget process can provide a useful forum on which to establish the strategic plan of the company. Whilst each year it will be necessary to produce a company fleet budget, it should be looked on as a means of providing realistic projections of cost for the year ahead and a base from which variables can be measured.

The budget can be seen by different departments in the company in a number of ways, although based on the same factual information:

(a) it can be a projected estimate of income and expenditure throughout the year divided into manageable sections that reflect the core business
(b) it can form part of the financial plan of the company, building on the data available to provide both financial and management information on which corporate decisions can be made
(c) it can provide a series of pictures showing the likely position of the company fleet at any given time in the year and overall provide a realistic projection of cost.

BENEFITS OF BUDGETARY CONTROL

The benefits of budgetary control are fourfold:

1. It provides standards against which to measure the efficiency of the different elements of the fleet.
2. It provides a base for comparing actual performance against an agreed objective.

3. It provides guidance for correcting objectives that are currently being missed.
4. It clarifies the role of the fleet management and its control, in financial terms.

TYPES OF BUDGET

There are three different types of budget:

1. The capital budget on which the forecast of the expenditure on capital assets can be made.
2. The operating budget which will cover all revenue and expenditure including administration and management costs.
3. The cash flow budget which will ensure there are funds available to meet the costs incurred.

For most company car fleets, there will be no revenue generated by the use of vehicles but if there were to be, a revenue budget would also be needed.

The Capital Budget

Fundamental information needed to compile a capital budget includes the number of vehicles to be replaced during the budget period and the additions or subtractions to the fleet; the type and cost of each vehicle; the timing of the fleet changes and the method of acquisition. Bear in mind that the method of acquisition will have an effect on the timing in terms of cost benefits for tax purposes and for a company with a large turnover of staff requiring cars, a policy needs to be established to ensure that there are never too many cars available.

Regular liaison with the personnel section will ensure the budgeted cost of the car fleet will at least be in line with the aspirations of the company manpower strategy.

The Operating Budget

This is essentially a "control of revenue" budget, rather than a control of capital. As such it is vital to ensure that all revenue costs are controlled (as far as possible) as close to the point where they are incurred.

Creating the budget

The five major areas need to be identified and the number of cost elements under these headings include the following.

Personnel
Number of people employed
Total salaries incurred
National Insurance and pension/other benefit costs
Uniforms or protective clothing

Establishment costs
Allocation of cost for area used
Rent charges
Heat, light, water

Vehicle costs
Depreciation or finance charges
Profit or loss on disposal
Licences and insurance
Repairs and maintenance
Fuel and oil
Replacement vehicles

Administration costs
Stationery and office equipment
Telephone rental and charges
Postage and miscellaneous expenses

Company overhead costs
Proportion of head office costs
Any allocation for other costs including training and education, recruitment, building repairs and employment legislation.

Cash Flow Budget

Estimation of cost

Having established the number and type of car and the method of acquisition, together with the timing of the programme, a cash flow profile can then be prepared.

When acquiring vehicles through one of the many financial methods available, beware of the commitment made at the time and ensure that the commitment is duly noted on any long-range budget forecasts that may cover more than the financial year under consideration. Certain financially attractive acquisition methods can be subject to heavy financial penalty if they do not run their complete term.

Phasing the budget

Phasing the budget (spreading the costs over the whole financial year in the order in which they are incurred) is necessary to ensure the cash flow demand can be met. If the fleet is large then even the cost of the road fund licence could have a significant effect on cash flow.

Similarly, with maintenance costs, although the lifetime repair and tyre costs can be forecast, it is virtually impossible to determine when the repairs will be carried out. Probably the only practical approach is to add up all the estimated costs in a given year and apportion the total cost equally over the year. Whilst this helps with the budgeting process, the cash required to take into account the actual spend will need to be accounted for separately.

FLEET MANAGEMENT

Whilst fleet management companies may provide a better control of cost than can be achieved in-house, the management of the budget and the incidence of when these costs will be incurred will not vary from the position of doing it in-house. The extra administration costs payable for such fleet management could be taken into account under the maintenance budget allocation where it would be hoped that the improved control, with a resulting reduction in maintenance costs, would more than adequately offset the increased charge for the management service.

Budget Review

If the process of budgeting is monitored accurately and the actual costs are shown clearly against the budget forecasts, it ought to be possible to investigate the reasons for over- or under-expenditure. Allowing for the dynamic nature of a company, certain key assumptions can be made regarding the reasons for variation and if these are accurately recorded, a forecast for the following year's budget will be more accurate. Ultimately, the budget process helps to set standards against which to operate. These standards, if monitored carefully, will enable the company to anticipate precisely where it is as regards revenue and capital expenditure at any time within its management accounts, enabling a precise forecast of the position at the end of the financial year.

Budget Forecasting

Although the company's operating costs will provide an accurate historical record, it is always useful to find time to compare in-house information on specific vehicles with that provided by the manufacturers and the comparisons in the trade journals. On a reciprocal trading basis, it may be possible to exchange information with similar sized fleets with similar operating criteria and review the local contract hire rates to ensure a range of comparisons. Note that problems can occur when using information from other than known sources if the quality of the data used is in any way suspect. Other companies' information may contain erroneous input or include items that another company deals with differently. It can also, of course, be out of date.

With so many factors in car fleet management dependent upon one another, it is worth remembering that much of the finance calculation is based on an assumption of residual value which is often a reflection of the make and model of the vehicle chosen. Most company car fleets are based on a limited range of vehicles all of which achieve a reasonable residual value of around 40% after three years. However, with the increasing demand for individual choice in the selection of vehicles, it is worth noting that the percentage value of the original price can vary across a range of vehicles from a low of 22% to a high of 67%. In part, it depends on how much the company is willing to invest in fleet cars as some of the more expensive models (such as Mercedes) depreciate far more slowly.

A current example of residual value is shown in Figure 2 which is plotted as a graph in Figure 3.

Perhaps the most significant feature of this data is the considerable first year depreciation of 61% to 64%, resulting in many companies looking to "nearly new" vehicles on purchase or contract hire to take advantage of the lower capital value.

Figure 2
Residual Value Projection — Table

MAKE/MODEL	"LIST" PRICE	PROJECTED VALUES							
		1 YEAR MILEAGE		2 YEARS MILEAGE		3 YEARS MILEAGE		4 YEARS MILEAGE	
		10K	20K	30K	60K	40K	60K	40K	60K
Ford Sierra 1.8 LX	11,646	7,000 61%	6,400 56%	5,500 48%	4,400 37%	4,450 39%	4,000 35%	4,125 36%	3,650 32%
Vauxhall Cavalier 1.6L 5 door	11,340	7,250 64%	6,800 60%	5,900 52%	4,550 40%	4,650 41%	4,200 37%	4,350 38%	3,850 34%

Figure 3
Residual Value Projection — Graph

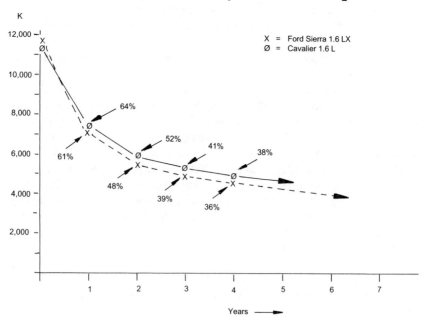

Chapter 5
ACQUISITION OF CARS

FUNDING CHOICES

For the new fleet manager there can appear to be a bewildering array of options when it comes to acquiring a car fleet. As Figure 4 (see overleaf) shows, these options can be quite quickly parcelled into manageable portions. Each option is clearly linked to a sub-option as shown. Many more complicated variations continue to be generated by the industry to overcome particular tax or operating problems but they can all be fitted into this simple diagram with little effort.

As with most aspects of car fleet management, it is worth going back to basics to establish a firm foundation on what the company precisely needs as this will enable the fleet manager more accurately to choose the range of "products" which meet that need.

STAGE 1

First, establish where the real need to acquire vehicles stems from. This sounds easy but company policy may well dictate the need and the choice on whether to purchase or rent can still be difficult to establish. Is the policy adopted by the company realistic? Perhaps it would be advisable to look at options and compare them both financially and professionally as a responsible manager — taking the overall "company" perspective.

Checklist

- Is there a company policy document that covers company cars? When was it produced?
- When was the situation last reviewed?

Figure 4
Fleet Acquisition Considerations

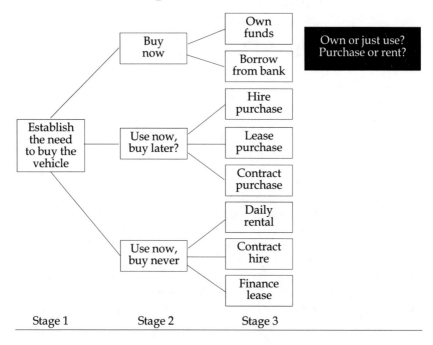

- Who reviewed the situation and can they provide more background on why the choice was made the way it was?
- Who uses the cars and why? Are they *essential* to everyone or part of the salary package?
- What mileage/fuel/maintenance cost records are there and is the cost information accessible and detailed?

It may not be possible to get complete answers to these questions but the attempt should be made before moving on to Stage 2.

STAGE 2

Company policy may dictate one of three options:

(a) to buy outright now
(b) to use now but delay purchase until sometime in the future
(c) to use now but never to buy.

Alternatively there may be no company policy. Either way, these options need to be considered in view of what will have been found at Stage 1. It is worth pointing out that some 60% of all companies operating fleet cars acquire at least some of their vehicles with their own cash. The reason for a cash purchase is probably the extreme flexibility it provides but there are some disadvantages in using cash which only further investigation will show by evaluating Stage 3. Having reviewed the overall policy, probably with the finance director, move on to Stage 3.

STAGE 3

Depending upon the earlier decisions, the "buy now" option results in two choices: either to use the company's own funds or to borrow from a bank or other lending institution. However, if the decision is to use the asset now but delay its purchase, there is the choice of hire-purchase or lease-purchase (which are explained in more detail below). Before that, bear in mind that all of these options lead to the next question — who will manage the vehicles?

If earlier decisions have led the company to decide that it does not want to buy the vehicles, then the choice is down to daily rental, contract hire, or a number of finance lease options. Whilst many aspects of control will be covered by daily rental or contract hire companies, there are other general management aspects that need to be considered. Stage 3 consists of a range of "products" currently available in the industry, each designed to meet specific company needs. The major acquisition products are detailed as follows.

Buy Now

Use own funds

This remains the most popular method of acquiring company cars with over 60% of companies currently buying their vehicles provided they have funds available. Payment is usually made to the dealer at the end of the month or when the car is delivered and part of the payment may be the trade-in value of the vehicle being replaced. The benefits are as follows:

(a) freedom of choice to change vehicle and/or supplier
(b) substantial discount for cash, particularly if there is no trade-in
(c) if funds are freely available and relatively cheap (low interest rate for borrowing) to the company, then the benefits of being able to make a quick decision, the flexibility of choice and the ease of acquisition can be very attractive.

The disadvantages, however, can outweigh the advantages and in brief they are summarised as follows:

(a) an excessive proportion of the company's funds may be tied up in vehicles
(b) most cars lose value each month (depreciation) and the more expensive cars *can* be worth half of their value within two years, however well they are looked after
(c) loss of interest on funds used to acquire cars and/or return on capital employed
(d) cash may be demanded for buying vehicles at a time which is inconvenient to the company. There may be more pressing priorities so planning for replacement vehicles may be difficult to achieve if there are restrictions on cash or credit
(e) risk regarding the residual value of the car on disposal is entirely the company's and the market may collapse.

Buying a car outright remains a popular method and can still be the most economical scheme for a company with a low rate of return on its capital employed. For an expanding company it can be shown that better use can be made of the company finances and other forms of vehicle acquisition such as contract hire or contract purchase can be more financially attractive.

Use Now — Buy Later

Borrow from a bank or finance house

This option is very similar to a company using its own funds except, obviously, rather more expensive. It retains all the advantages of outright purchase but as far as the disadvantages are concerned, it absorbs one of the possible sources of credit the company might otherwise need for other purposes. Also the interest rate charged by the lender may significantly alter the balance in favour of other acquisition options. If borrowing money to acquire cars is a necessity, then all other options should first be evaluated before a decision is taken.

Hire purchase
(This also operates under the name of lease purchase or deferred purchase.)

Hire purchase is one of the choices under the "Use Now but Buy Later" headings and is simply a method of buying the car by instalment payments. There is a minimum deposit to be paid and some agreed time in which the total amount agreed has to be repaid. An "option to purchase" or "title" fee is paid on the last instalment which transfers ownership from the finance house to the company. Payment of the balance due between the deposit and the title fee is made over the period of the agreement which can be up to five years (but note that title or ownership does not pass until completion of the agreement).

The advantages of hire purchase can be itemised as follows:

(a) initial deposit (often misleadingly called the "rental payment") is usually much smaller than the cash payment required for outright purchase
(b) hire purchase, put simply, is buying a car by monthly instalments which are paid out of revenue, not capital
(c) tax relief is available (see *Company Tax Treatment* later in this chapter).

The disadvantages can be summarised as follows:

(a) the interest rate charged by finance houses (fixed or variable) is often high and the deposit required is controlled by Government rules
(b) the agreement is for a fixed period of time so flexibility is lost. The car cannot be sold until the company has gained title to it and to do this earlier than was predetermined will require a settlement fee to be paid.

Lease purchase
(Variations)

Lease purchase is a further product of the finance houses and is devised to overcome some of the limitations of the Hire Purchase Act which requires a minimum deposit before goods can be delivered. The lease purchase agreements are based on a low initial payment together with the monthly payments to take into account the anticipated residual value of the car when the agreement ends. The company then pays a cash sum equal to this value (which can be quite substantial) to complete the purchase. This is usually referred to as a balloon payment. Additionally, the company pays an option to purchase or "title" fee in order to gain legal title to the asset. The payments, often misleadingly called "rentals", do not attract VAT.

Contract purchase

This is a further "product" of the industry to overcome some specific tax areas that other products do not cover. (The significance as to whether contract purchase is right for a company will depend upon the company's position on VAT and corporation tax and although a technically interesting product, contract purchase has only managed to attract 4% of the market to date.) A contract purchase deal is usually written as a conditional sale agreement. When offered by a fleet management company, a vehicle is acquired by them and they "sell" it to a customer. After the last payment is made, the customer company may exercise an option for the fleet management company to dispose of the vehicle on its behalf, or it can retain the vehicle and dispose of it itself.

As the deal is considered as a financing agreement only, the capital and finance elements of the rental do not attract value added tax. The fact that VAT is not payable is particularly attractive to organisations who are unable to recover VAT in full such as banks, building societies, property and insurance companies.

As VAT is not payable on monthly depreciation and finance payments, significant savings can be made. Any additional service provided such as fixed price maintenance built into the agreement *is* subject to VAT.

Capital allowances for corporation tax purposes are currently restricted in respect of cars costing over £12,000 and there can be a permanent loss of relief for expensive cars if they are acquired through normal lease or contract hire arrangements (see Company Tax Treatment below). To overcome this disadvantage, the industry has developed contract purchase schemes which are structured to obtain the maximum tax benefit. To achieve this, the company must have beneficial ownership of the vehicle during the period that this financing arrangement is in place. As the transaction is a conditional sale, this condition is met and therefore the company is entitled to the same writing-down allowances as if it were acquiring the car with its own funds. When a contract purchase arrangement is made the disposal arrangement for the vehicle has to be confirmed at the time and for this there are two procedures, both of which will affect the accounting treatment under Statement of Standard Accounting Practice SSAP 21 (see paragraphs on SSAP 21 later in this chapter).

(a) *Agency Agreement*:
Under an agency agreement the fleet management company (usually) disposes of the car on the customer company's behalf for a small agency fee. The cash generated from the sale of the vehicle, less commission, is returned to the customer. In this way, as the customer company participates fully in the risks and rewards of ownership, the contract must be shown on its balance sheet.

(b) *Option Agreement*:
Right at the start of the agreement with the fleet management company, a resale value of the car, guaranteed by the fleet management company and payable on the termination of the agreement, is fixed and the customer company will have the option to call for this price. This value will be set at a prescribed number of months and at a pre-agreed mileage. If it is the customer company's contention with its tax inspector that it will always exercise the option, then, *provided its auditors agree*, the agreement may be considered to be an operating lease for the purposes of SSAP 21 and the car can therefore be "off balance sheet".

As before, if any additional services such as fixed price maintenance are incorporated into the monthly repayments, then while the costs will be allowable as operating expenses, VAT will be applied.

Contract purchase is generally more cost effective than contract hire for companies which are unable to off-set their VAT payments in full as previously identified.

It is also more cost effective in the case of expensive or executive vehicles costing over £12,000, although the actual point where contract purchase becomes more attractive than contract hire is somewhat confused by the loss of manufacturers' bonuses (internal "discounts" given to contract hire companies) and capital allowances available to the fleet management company.

Use Now — Buy Never

Daily rental

It has been found by many fleet operators that, depending on the need for company vehicles, it can be cost effective to undertake a contract with a daily rental company to provide vehicles as and when they are needed. Whilst regional variations and the size of the company demand will significantly influence the break-even point, it seems clear that if the use of the company car is only for company business and that need is no more than (say) 100 days a year, then it is likely that daily rental would be a viable option. This may not be acceptable as far as the user is concerned but from a company point of view, it could be the most cost-effective.

The advantages are:

(a) a wide choice of vehicles suitable for any occasion, "instantly" available
(b) vehicles can be delivered to the point of collection most convenient for the user
(c) no significant fleet management administration is required
(d) cars are only paid for as and when they are needed (an advantage over the pool car system).

The disadvantages are:

(a) the lack of convenience of having own car parked in the garage
(b) potentially higher cost if the need for the car exceeds expectation
(c) reliance on the "quality of service" from the rental company.

Contract hire

This form of use of an asset without ownership is growing in popularity. Over 20% of fleet cars are acquired in this way and the use of contract hire is still growing. Essentially an operating lease, the risks and rewards of owning the vehicle remain with the contract hire company and, consequently, one of the main benefits of contract hire is that the rental payments are paid from the revenue account of the customer company and have no effect on the capital or balance sheet of the customer company. At no time does the customer company own the vehicle.

The contract hire company takes all the risk on the residual value of the vehicle but covers that risk with a written agreement which includes specific total mileage to be achieved in a given time. Penalties for excess mileage are likely to be built into the contract and if contracts are extended there may be penalties to pay as the contract hire company will have anticipated a residual value which may be significantly different over a longer time frame.

A benefit of a contract hire arrangement is that all aspects remain negotiable and depending upon the size of the fleet, there can be significant room for negotiation.

There are two primary benefits for a company:

(a) it has no involvement in ownership, therefore fluctuations in residual value are not its problem
(b) under most circumstances, the finance interest rate applied for the period of the contract remains fixed.

The contract hire agreement generally allows, for a predetermined monthly cost, the use of a vehicle within a given set of conditions. It is only when these conditions are exceeded or the customer company wishes to change them after the contract has been signed that particular care needs to be taken on the renegotiation. Contract hire is particularly useful for budgeting and cash flow projection purposes and the choice of suppliers is very wide.

The majority of contract hire agreements incorporate a fixed price maintenance package which often includes servicing, repairs, tyres, batteries and exhausts. The road fund licence cost is usually included in the contract as is roadside assistance (AA, RAC, etc) and the cost of a replacement vehicle in the event of mechanical breakdown or accident damage (within agreed parameters).

Alternatively, fleet management companies will undertake the physical management of the vehicles and provide the link between the vehicles and the garages that maintain them but fleet cars are generally a very emotive subject. Although responsibility for certain aspects of car management control is delegated to the contract hire company, nonetheless a great deal of co-ordinating management responsibility still rests with the customer company's fleet manager.

For daily rental and contract hire, much of the management control offered is clearly laid down in the terms and conditions agreed between supplier and user. For large fleet operators, many of these contract conditions are very negotiable. It is essential, however, that the customer company retains overall control by stipulating the conditions under which the vehicles are provided and that they meet its needs.

Finance leases

The car fleet industry has a huge range of finance lease products available to customer companies ("lessees") and the product is sufficiently robust to enable the specialist leasing companies ("lessors") to "tailor" lease contracts to individual customers' tax, cash flow and other needs. Broadly though, there are three types of finance lease available:

- "balloon"
- "closed-end"
- "open-end".

These are described briefly below.

Balloon leases

Under a balloon lease an initial assumption is made about the residual value of the car at the end of the lease period. A final ("balloon") payment is due at the end of the lease period which is equal to this residual value. In theory the final balloon payment should be matched by the sale proceeds of the car and incur no additional liability for the lessee. The monthly rental payments under a balloon lease are lower than under a lease contract which repays the whole capital cost and the lessee pays only for the lessor's financing charge and the projected depreciation of the car.

The advantages of balloon leases are:

(a) lower rental payments
(b) final balloon payment can be met from sale proceeds of car
(c) attractive when inflation rates and residual values are high.

The disadvantage of balloon leases is the risk that the initial residual value calculation will be wrong and the sale proceeds from the car will not match the final balloon payment, in which case the lessee will have to fund the shortfall.

Therefore low rental payments, high residual value contracts should be avoided.

Closed-end leases

Under a closed-end lease, the lessee makes rental payments over the lease period (usually two to five years) which are sufficient to cover both the full cost of the car over the rental period and the lessor's financing charge. In other words, the car is fully amortised over the lease period and for this reason these leases are sometimes known alternatively as "fully amortised" leases. On expiry of the lease, the lessee receives a rebate of rentals to reflect the sale proceeds of the car.

The advantages of closed-end leases are:

(a) the lessee knows in advance the full extent of his or her commitment and does not have to make additional payments if the residual value of the car at the end of the lease is lower than initially anticipated
(b) the lessee receives a rebate of the sale value of the car at the end of the lease period.

The disadvantages of closed-end leases are:

(a) lack of flexibility
(b) where rebate of sale value of car at the end of the period is less than 100%, this constitutes a hidden and perhaps significant interest charge
(c) higher monthly rentals than under balloon leases.

Open-end leases

This term is now seldom used but reflects a rather more flexible arrangement than a standard closed-end lease. In particular, the lessee is given more choice about when to end the lease. The most common form of open-end lease is where an option is given to the lessee to continue with the lease into a secondary leasing period after expiry of the primary leasing period. The level of rentals payable in the secondary leasing period are merely nominal because the car has been fully amortised during the primary leasing period.

ACCOUNTING TREATMENT

It is important that fleet managers should be aware of the differing accounting treatments which apply to the various forms of financing the car fleet. Whilst it is relatively obvious that where a company buys a car, it should appear as an asset in the company's balance sheet with any continuing financing obligation as a countervailing liability, the accounting treatment of the various forms of lease rentals may be less clear to the layman. Fortunately, the position has largely been clarified by the publication of Statement of Standard Accounting Practice 21.

Statement of Standard Accounting Practice 21

Leases can be classified into two generic categories: finance leases and operating leases.

Under a finance lease, a customer company ("lessee") makes payments to a specialist leasing company ("lessor") to cover both the full cost of the car and a profit margin on the funds provided by the lessor. Although legal title to the car remains with the lessor, the normal risks and rewards associated with ownership are passed under the lease contract to the lessee so that the lessee is in substantially the same position as he or she would have been had he or she purchased the car. For example, the residual value of the car at the end of the leasing period will fall to the lessee's account so that he or she will make a profit if the residual value is higher than anticipated and suffer a loss if the residual value is lower than anticipated.

Under an operating lease, a lessee makes payments to a lessor for the temporary hire of the car normally for a shorter period than the full economic life of the car. In this case the lessor retains most or all of the risks and rewards of ownership. Contract hire arrangements constitute a typical operating lease structure.

SSAP 21 requires that a finance lease must be capitalised by a lessee company to reflect both the purchase of the rights to use and enjoy the car and the corresponding obligation to make future lease payments. The car must therefore appear in the lessee company's balance sheet under the "leased assets" category. Lease payments are accounted for as partly reducing the future obligation (and so do not appear in the profit and loss account) and partly as ongoing financing costs (which *do* appear in the profit and loss account as interest payments). The attributed capital value of the car lease must be depreciated over the shorter of the term of the lease and the projected useful life of the car. The depreciation rate should be set so that the depreciated value at the end of the lease term equals as nearly as may be the assumed residual value of the car.

SSAP 21 permits an operating lease to be reflected only in the lessee company's annual profit and loss account on the basis of the annual rental payments made in the course of the relevant financial year.

COMPANY TAX TREATMENT

Purchase

Where companies purchase cars either directly or through other purchase options such as hire purchase, contract purchase or lease purchase, they can claim capital allowances on the acquisition costs of each car. Although companies cannot reclaim the VAT payable on the acquisition of a new car, the VAT element can be included when calculating the acquisition cost of each car for the purposes of claiming capital allowances.

Cars costing up to £12,000

As from 10 March 1992, capital allowances can be claimed on cars costing £12,000 or under, broadly at a rate of 25% pa on a "reducing balance" basis. This "reducing balance" basis is illustrated in the table below for a car having a total acquisition cost of £10,000.

Prior to 10 March 1992, the maximum value car eligible for full capital allowances, total cost was £8,000.

Expenditure on cars costing up to £12,000 for cars purchased on 10 March 1992 or after and on cars costing up to £8,000 if acquired prior to that date should be kept in a separate pool of expenditure. The sales proceeds of any cars sold should be deducted from this pool of expenditure and the 25% capital allowance is calculated on the net balance remaining.

Year	Opening Balance £	Reducing Balance £	25% Capital Allowance £
1	10,000		2,500
2		7,500	1,875
3		5,625	1,406
4		4,219	1,055
5		3,164	

Cars costing over £12,000

Where a company acquires cars with a total cost of over £12,000 (£8,000 if acquired before 10 March 1992), they must be separately identified. This is because the maximum annual capital allowance for each company car is £3,000 (£2,000 for cars acquired prior to 10 March 1992).

When these more expensive cars are sold, a "balancing allowance" calculation has to be made. Where the car is sold for less than its written down value for corporation tax purposes, the company will receive an allowance equal to the shortfall. If the car is sold at a value higher than its written down value for corporation tax purposes, the company will face a balancing charge equivalent to the amount to which the sale proceeds exceed the written down value. Consequently the total capital allowances available on more expensive cars will through the medium of either a balancing allowance or balancing charge equate to the actual amount of depreciation, if any, over the period.

It follows therefore that the impact on the restriction on the annual amount of capital allowances available on more expensive company cars operates only to defer capital allowances and does not permanently confiscate them.

Car accessories

Those car accessories which are fitted prior to delivery of the car are regarded as increasing the acquisition cost of the car for the purposes of capital allowances. Where accessories are purchased and fitted after delivery of the car, they are to be treated separately and would normally attract their own independent capital allowances.

Capital gains tax

Since cars are specifically exempt from capital gains tax, gains and losses on the sale of company cars have no impact for capital gains tax purposes.

Loan interest

Interest paid on loans to fund the purchase of company cars will normally be a deductible expense for corporation tax purposes.

Finance Leases

The leasing company is entitled to the same capital allowances described above for a company purchaser. The corporation tax treatment for the corporate employer who enters into the leasing agreement depends entirely upon whether the leased cars had an acquisition cost of up to £12,000 or over £12,000 (£8,000 prior to 10 March 1992).

Where lease rentals are paid in respect of cars having an initial acquisition cost of up to £12,000 (£8,000 prior to 10 March 1992), the lease rentals paid are fully deductible for corporation tax purposes.

Where lease rentals are paid in respect of cars costing over £12,000 (£8,000 prior to 10 March 1992), a proportion of the lease rentals are disallowed on the basis of the formula set out below.

$$\frac{12{,}000 \text{ plus } \frac{1}{2}(P - 12{,}000)}{P}$$

Where P is the acquisition cost of the car.

Using this formula, the proportion of lease rental costs which will be allowed for more expensive cars is set out in the table below.

Acquisition Price (£)	Lease Rental Allowed (%)
40,000	65·00
30,000	70·00
25,000	74·00
20,000	80·00
17,500	84·29
15,000	90·00
14,000	92·86
13,000	96·15

Contract Hire

Cars under contract hire arrangements are subject to the same tax regime for corporation tax purposes as cars under a finance lease. This means that cars costing over £12,000 will have a proportion of their hire costs disallowed. For such more expensive cars, the basic hire charge (which should have a proportion disallowed) should be clearly distinguished from the maintenance charge (which should not have a proportion disallowed).

Chapter 6
CAR COST CONTROL

Vehicle fleet costs are a significant overhead in any organisation, and in a time of recession and redundancy fleet managers are asked to scrutinise more closely the costs associated with running all company vehicles.

Accurate, *comprehensive* records provide part of the answer and many large fleets have established computer-based systems that process the raw data of fleet management and can be designed to overcome, or at least minimise, human error in input.

Information that can be provided from this data can be extremely comprehensive but the *interpretation* of this information and its use and application is often not complete.

Standards can be incorporated into the programmes that are time or mileage relevant in order that variations to these standards can be quickly identified and investigated. However, the willingness or authority of in-house fleet management to act on the information can be lacking.

The fleet manager is often challenged with the task of achieving *front-end discounts* on the supply of new vehicles rather than on the *whole-life vehicle cost* which is not just finance led but takes into account all the backup services necessary for a truly efficient system.

The employees responsible to the fleet manager for implementing the scheme may not be *fully trained* and yet it must be clear that professional fleet management, whether from outside specialists or dedicated in-house personnel, ultimately provides a company with a method of maintaining tight control over its vehicle fleet running costs.

There are many specialist *fleet management* companies established that can offer a full range of services which can be used individually to supplement in-house resources or can be used as a total package.

If the professional fleet manager, either in-house or bought-in, is to maintain tight cost control of the fleet, then he or she has to be *adaptable to change* and capable of responding to the increasingly sophisticated products of the industry that can supply the fleet managers with the range of services they require.

Whilst many company fleet managers may consider that the fleet management companies pose a threat to their wellbeing, this should not necessarily be the case as the special needs of each company will still require the *professional expertise* of individuals who can relate the products and services to the core business of their company.

WHOLE-LIFE COSTS

Although many companies consider the capital value of the car to be the most important feature, it is worth realising that all vehicles vary in terms of their operating costs and in particular the residual value will have a considerable impact on the total cost of the vehicle to the company.

Figure 5 below gives a diagrammatic representation of whole life costs which start with the purchase price, to which is added the operating cost and *from* which is subtracted the residual value, to arrive at the whole-life cost.

Figure 5
Whole-Life Costs

WHOLE-LIFE COSTS			
Purchase Price	Operating Cost	Residual Value	Whole-Life Cost
Quality image Precise specification "Real" quality Durability Driver acceptability	Fuel consumption Maintenance cost Time "off road" Tyre wear/cost Specification right!	Quality image Resaleability Service record Condition Recondition	

PURCHASE PRICE

Over 60% of a vehicle's original value can be lost in the first 12 months, even with relatively low mileage. There are exceptions to this rule which apply to more expensive vehicles which are produced in small numbers. These often remain in high demand, with much lower depreciation in the first year and indeed throughout their whole life. Whilst it could therefore be argued that all company vehicles should meet the objective of having a high residual value, the capital cost involved would be prohibitive in practice.

The decision on the vehicle purchased must depend initially on the company policy with regard to the range of vehicles that can be chosen. Within this range, each specification should create a quality image, both to the original driver and to the subsequent purchaser of the vehicle (say) three years hence.

The precise specification of each model will need to be carefully considered since exotic expensive options may not find favour with the end user in three years' time and therefore will provide no additional value to the residual value of the car.

Another consideration is the *"real" quality* of the vehicle in the eyes of the professional car dealer. Whilst some vehicles will be seen by their new owners to be the latest thing in technical excellence, they may not be perceived as such by professional motor engineers due, perhaps, to some fundamental production or material faults that have yet to be fully resolved by the manufacturer.

This introduces the subject of *durability*, where the link between a high purchase price and vehicle durability would be assumed to be very close. This may not be the case for some of the reasons already mentioned and overall durability may not be as realistic as the original encouraging sales information might lead you to believe!

Finally, consider *driver acceptability* where the combination of acceleration, good road holding, excellent braking and driver comfort will have impact both on the initial purchaser and the subsequent buyer. A car with poor steering geometry will not significantly improve on this feature no matter how attractive the vehicle looks or how much is spent on maintenance.

OPERATING COST

The total life of the vehicle and all costs incurred after the initial capital cost has been paid for the vehicle's acquisition need to be anticipated and considered.

Fuel consumption will be high on the agenda, starting with consideration of fuel type. For high mileage vehicles, the benefits of diesel will be very important. Whilst many company car drivers are reluctant to switch to diesel, the price will be attractive (particularly if they pay for their private fuel use) and the reliability of diesel under adverse weather conditions will be an added bonus. The disadvantage of the smell has been overcome to some extent by the suppliers adding deodorisers and more significantly by the fuel stations providing cleaner facilities for diesel handling. Plastic gloves are now readily available in most garage forecourts when using self-service (diesel doesn't evaporate as quickly as petrol and lingers on the hands if one is in direct contact). The big advantage is the relatively few times the garage has to be visited to refill the tank as a tankful of diesel will probably give well over 500 miles of motoring.

With most cars fuel economy and environmental responsibility mean switching to unleaded petrol and then being fitted with a catalytic converter. The costs are lower than for leaded petrol at the pump, but the catalytic converter requires a higher fuel burn to create the same amount of energy so some of the financial advantage is lost in the higher fuel consumption.

The cost of vehicle *maintenance* will vary significantly. It will depend on the type of fuel used and the nature and size of the engine specified, as well as all the extra components such as air conditioning which may be fitted. A wealth of information is available on maintenance costs from trade magazines and manufacturers' details, all of which should be used in comparison with the company's own data from its own fleet management control system. Some vehicles require more maintenance than others or break down more frequently. The time off the road when replacement vehicles are required can add considerably to the whole-life cost.

Tyre wear and tyre specification is a very broad subject but, briefly, it is important to ensure that full manufacturers' discounts are obtained on any new tyres fitted and that the tyres are not over-specified for the performance of the vehicle in question. High-speed tyres do not make the vehicle go faster! Naturally, care in driving techniques and in maintaining the correct pressure and tracking of the tyres is essential for long life. Excessive tyre wear is often the first obvious sign of abuse of a vehicle by a company car driver.

The specification must match the use to which the car will be put as this will have an effect on the operating cost. Heavy equipment being moved around in the boot of a light family saloon can create great damage very quickly and long motorway driving (as a regular feature of a driver's need) will probably not be enhanced by a high revving small engine working near its maximum output.

RESIDUAL VALUE

The company car is likely to be sold eventually to a private motorist. When specifying the vehicle for company use, it is worth bearing this in mind if the maximum resale value is to be achieved. For instance, the *image* of quality for the private buyer will perhaps be a reflection of a different set of objectives than that of a company car owner so over-specifying a company car can be a disadvantage.

The resale value will reflect the perception that the industry has of that make and model as well as the private buyer's view on whether the need is for a high quality interior (for example) or a utilitarian vehicle for work. Colour may have an impact on a vehicle's ability to be sold; for reasons that cannot easily be explained, red is still the most popular colour.

The *service record* of the vehicle is vital. This will give the potential buyer not only details of what was done to the vehicle before they acquire it but also the fact that it was serviced regularly within the manufacturer's guidelines. This will help to confirm that the mileage given is genuine since most garages put the mileage on a service document when the car is serviced. This can help to prevent "clocking" (a means by which the residual value can be enhanced by indicating the car has travelled considerably less miles than has actually been the case).

The *condition* of the vehicle will have an impact on its resale value and it is worth considering surcharging company drivers for loss of value through abuse of their vehicles. Certain deterioration cannot be avoided (eg to the driver's seat) but rips and tears to other upholstery, badly soiled carpets and upholstery with ingrained dirt from dogs, etc may only be removable by professional valeting.

Other more expensive mechanical work, thought necessary to bring a car up to full specification for resale, has to be considered very carefully as it is easy to overspend on mechanical refinements which will not significantly improve the residual value. Reconditioning needs to be approached cautiously.

INSURANCE

As a result of increasing theft and the high cost of repair, along with the relatively high incidence of accidents, insurance premiums are increasing dramatically. The only way the fleet manager can reduce the cost is to improve controls and spot problem areas early. This means a company must give clear instructions to company car drivers to ensure they are clear on their responsibilities and on who they may allow to drive the company car.

A brief checklist of items to consider may help to focus on the major issues that are particularly relevant.

1. Check documentation of all company car drivers and at least once a year physically check all drivers' licences for endorsements.
2. Ensure that the accident reporting system works well and is well documented in the company handbook.
3. Make safe driving part of the company's safety policy.
4. Make it worthwhile for drivers to keep insurance premiums low by developing an incentive *Driver Award Scheme* but balance this with a penalty if the driver is blameworthy. The premium excess on an insurance policy, paid directly by the driver at fault, could encourage others to drive with more care.
5. Encourage membership of the Institute of Advanced Motorists. This will help reduce premiums and may be worth considering as a company incentive package.
6. Consider making driver training an integral part of conferences or seminars and not necessarily a punishment for accident-prone individuals.
7. Fit approved alarm systems to all company cars. Some insurance companies will offer reductions in premium as a result.
8. Be aware of diesel discounts. Following new insurance groupings, there are incentives for diesel vehicles and certain categories of diesel drivers.
9. Keep a look out for new products in the industry where insurance is included in the package. Some companies will include insurance, maintenance and finance in the provision of a new vehicle, and this may be a cheaper combination than you can achieve for yourself.
10. Be aware of the choice of vehicles for company car drivers. Some vehicles attract a significant insurance premium either because they are at high risk from theft or are very expensive to repair.
11. Professionally manage the fleet and provide clear guidelines for drivers. Generate a company car driver's handbook that will give company policy direction and monitor all major activities to ensure that you are in control of the costs. For example, accident management can now be purchased as a further product of the car fleet industry.

It is worth investigating the options available on insurance and seeking alternative sources of supply. A good broker will probably manage this aspect but if the company runs a large fleet of vehicles, it is possible to insure directly.

REPLACEMENT CYCLES

Vehicles will eventually need replacing but it can sometimes be difficult to determine just when this should be. In practice, this will often be determined by a company policy which may be based on the following list of "alternatives".

ALTERNATIVE REPLACEMENT POLICIES

1. Pure guess.
2. Inspired guess.
3. Look out of the window and copy the company next door.
4. "It's just collapsed, I must replace it."
5. "I always have a new car before I go on holiday."
6. "I must have the latest registration plate."
7. "There's an extra 10% discount on offer if I buy before the end of the month."
8. The thought-through rational economic approach.

The figure overleaf gives an example of the rational economic approach that a company can develop based on its own records.

If a plot is made (against time and/or mileage) of the regular areas of cost for the vehicle such as its servicing, tyre replacement, breakdown costs, replacement unit, road fund tax, administration costs and any other defined areas, "windows of opportunity" will appear.

The windows are simply times in a vehicle's life when costs have already been incurred but there is a gap before the next round of costs will start. These windows commonly occur after one year or 25,000 miles or three years and 60,000 miles. This is why many companies tend to choose these times in determining their replacement policy. However, it is better to base a replacement policy on a company's own records. They are more precise and specific to the fleet. A company may discover that their first "gap" is at 14 months and the second gap appears at 3.5 years.

If the point of "optimal" turnover is passed and a subsequent decision is then made to retain the vehicle, it is likely that the costs will escalate significantly. These costs are very difficult to control since they can, in part, be a result of the reaction from the employee to not receiving a new car. All of a sudden, tyres that were perfectly adequate *have* to be replaced. Bodywork paint chips which could have been suffered *have* to be resprayed. Minor squeaks or whines in the engine or transmission or suspension *have* to be attended to. The fleet manager has little choice but to concur, despite severe budget constraints, in order to "keep the peace".

There is no easy solution to this problem since, in these recessionary times, vehicles are kept for longer than would ideally be the case. The significance of retaining cars, particularly for high mileage users, cannot be overemphasised in terms of the much higher costs that can be incurred — a point to be very carefully monitored.

Figure 6
Vehicle Cost Replacement Cycles
"Windows" of Opportunity

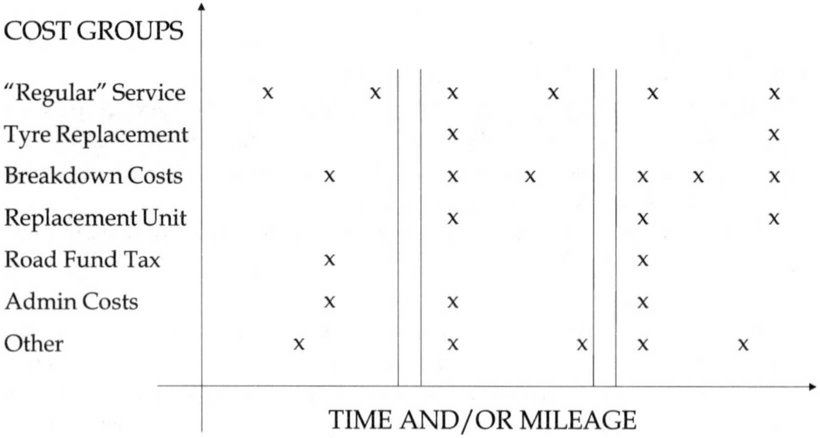

COST GROUPS								
"Regular" Service	x		x	x	x	x	x	x
Tyre Replacement				x				x
Breakdown Costs		x		x	x		x	x	x
Replacement Unit				x		x		x
Road Fund Tax		x				x		
Admin Costs		x		x		x		
Other	x			x		x	x	x

TIME AND/OR MILEAGE

Chapter 7
DISPOSAL OF CARS

Fleet car management is considered less attractive an occupation when it comes to vehicle disposal but while great effort will be made by those who purchase their cars to negotiate front-end discounts, far more can be saved by also concentrating on ensuring the vehicle achieves its maximum residual value. This can be less than 40% of its original purchase price.

Many hundreds of pounds can be gained on the sale of a car in good condition where maintenance records are clearly shown, the car has been valeted professionally and all minor damage eliminated properly. Before considering at length the curative solutions, one should first consider the preventive options to ensure all vehicles returned by company car drivers are in an acceptable, previously agreed condition. This could be achieved by the enforcement of penalties. Regrettably, as company cars are seen as a rather emotive subject, without senior management support, fleet managers may be hard pressed to introduce all the controls they would wish.

Timing

If the disposal of fleet cars can be delayed to certain months of the year, a higher residual value will be achieved. Although in recent years there has been a less noticeable seasonal demand, it still seems true that March and April show an increased interest by private buyers and there still remains a shortage in supply of cars generally in June and July as both company and private buyers wait until August to replace their vehicles.

December, January and February have historically been poor months to gain best residual values. Christmas is not a popular time for buying cars and bad weather in January and February discourages people from looking.

For the manager of a large fleet, many of the major auction houses are now offering a comprehensive service on estimating more precisely what particular models are likely to fetch when they are made available. However, as a general rule, if the disposal of cars can be delayed by using them in some other manner within the fleet in order to hit the peak time for sale, the chances are that better values will be achieved.

The price obtained for a used car is subject to a number of factors which include:

- the number of similar vehicles offered for sale at the same time
- the demand for the vehicle
- its condition
- its colour
- the locality
- the time of year
- the weather.

This list is in no particular order and each prospective buyer will consider these factors quite differently.

DISPOSAL OPTIONS

There are several methods of disposal and all have their uses. Even public bodies who must only dispose of vehicles through public auction should be aware of the alternatives.

The main choices of disposal routes are:

1. Auctions
2. Used car traders
3. Sales to staff
4. Part exchange for a new car
5. Sales to the general public.

There are strengths and weaknesses in all these systems and perhaps a combination of methods is the best policy to adopt, depending on the volume of vehicles being sold.

AUCTIONS

Large fleets would normally clear most of their stock this way but companies would be advised to negotiate carefully on the terms of sale to ensure the best price for the range of services being offered. Larger fleets would probably negotiate a single charge that covered all vehicles whereas a smaller operator would probably rely on a sliding scale related to the actual prices achieved.

The services offered include collection, cleaning or preparation, and any service history reports that may be required. A good partnership between the fleet manager and an auction house can result in the provision of accurate advice on setting realistic reserve prices, bearing in mind that a reserve price set may bear no relationship to the written-down value on the company books. Special car sales are organised for particular makes of vehicle and certain types of vehicle sell better in certain areas so moving cars around the country through the auction house may be justified.

An auction firm does not purchase the car but only acts as an agent for the company, so it is essential that the terms of business are clearly understood. The completion of the entry form by the company gives authority to the auction company to sell the car to the highest bidder, subject to the offer price being equal to or greater than any reserve price stated on the entry form. Acceptance of an offer is confirmed by the auctioneer's hammer and, in law, this constitutes a binding contract between the seller and the buyer. In practice, there is a certain element of tolerance after this time in the event that major faults are found.

The buyer will pay the auction company the price offered and the company receives that sum, less the previously agreed charges. If no reserve price is set, the auctioneer has the right to sell the car to the highest bidder and the owner has no right of redress if the price obtained is lower than expected. Normally, however, no deal will be struck, even when there is no reserve price, without some reference to the owner.

Often auctions offer no warranty directly but separate warranty insurance cover can generally be arranged under certain conditions.

USED CAR TRADERS

A high proportion of cars sold at auction are subsequently resold many times within the trade before retailing to the general public. The price realised at auction, therefore, is lower than could possibly be achieved through a trader. The advantages of using a trader are:

(a) Cars can be sold as soon as they become available.
(b) They are removed immediately after sale.

(c) The agreed price is paid immediately.
(d) There are no fees payable.
(e) There is less paperwork.
(f) The legal responsibilities for roadworthiness are transferred to the trader.

The disadvantages are:

(a) The company must ensure that prices offered are in line with current trade prices.
(b) Traders tend to select the best cars.
(c) The company may have to move the car to the point of sale.
(d) Internal auditors are suspicious of car traders.
(e) It is difficult to monitor the transaction if under the control of subordinate staff.

Establishing contact with a reputable trader could be financially beneficial but care should be taken to release vehicles only when the cheque has cleared or unless other confirmed arrangements have been made. It is also worth ensuring that cheques are the only acceptable payment for the cars purchased and that a sales invoice through the company is generated giving precise details of the vehicle. The value of the road fund tax should either be taken into account in the negotiation or removed and reclaimed.

SALES TO STAFF

This can produce good financial results and can also generate good relations with employees. Normally the risk of serious mechanical failure is fairly low. There is however, the possibility of over-maintenance of the vehicle prior to being acquired by employees, particularly if it was their own vehicle, which results in extra cost to the employer. The risk of this happening is higher when there is no adequate maintenance control in place, but even where it is, inadequate management supervision may condone the act.

Agreeing a price can be a problem if the individual can obtain details of the written-down value and negotiate from that stance. Companies tend to be overgenerous in settlement as an expression of goodwill. The Inland Revenue can react strongly where a car has been sold for a price lower than any reasonable definition of current trade value and the employee could be taxed for this as a benefit-in-kind.

Because of the possible pitfalls that can occur, it is generally recommended that the car is sold through auction and employees are informed of the place, date and time so that bids can be made on their behalf, either personally or by telephone. Under these circumstances, employees get the opportunity to purchase cars at a reasonable price and the company fulfils all its obligations.

PART EXCHANGE FOR NEW CAR

This may apparently achieve a higher price than the other methods but usually this is at the expense of discounts on the new vehicle and more administration on each vehicle involved.

SALES TO THE GENERAL PUBLIC

Retailing cars directly to the general public can, naturally, be lucrative but its overall benefit to the fleet is doubtful. The secondhand car market is now relatively sophisticated, there are many competitors and several areas of legal and financial risk. Along with advertising costs, preparation of vehicles and staff time, there are the problems of insurance on test drives, the possibility of the legal implications of the Sale of Goods Act, and the creditworthiness of the buyer.

PRE-DISPOSAL PREPARATION

The recorded mileage and age are still the major value indicators of a used car, but improved speed of sale and maximised value can be achieved by spending time on preparing the car before it is offered for sale. An appropriate valet to the car will be highly cost-effective for fleet owners. Specialist companies or local garages will normally provide a fully comprehensive service for as little as £15 which can add hundreds of pounds to the resale value and, perhaps more importantly, clear the car quickly in preference to others. A car awaiting disposal can cost at least £5 a week in interest charges alone, so improving speed of sale is well worth the effort. The decision on whether to repair body and mechanical damage is more complex. Significant accident body damage should be attended to but minor blemishes are probably best left untouched.

Mechanical defects are more difficult to account for and significant component failure should, by law, be repaired. However, professional advice should be sought before spending large sums of money on engine overhaul, as the cost of the overhaul could be a large proportion of the value of the vehicle in its repaired state.

As stated earlier, the major factor in ensuring a vehicle is in good condition at disposal time is to make the company car driver responsible for the state of the vehicle from the beginning. A number of companies have introduced both incentive and penalty schemes for drivers with prizes for the best kept vehicles and penalties for those that cost money on disposal.

There are two issues worth remembering when it comes to disposing of vehicles.

1. The vehicle belongs to the company until such time as it is disposed of and as long as it is owned, it is costing money. Speed of disposal is therefore essential.
2. The car is also a company asset until its realised value is in the company's account. Selling cars on the company's behalf by a third party, where there can be a delay in remitting funds, should be avoided.

Chapter 8
TAXATION OF COMPANY CARS

COMPANY CARS — THE PRESENT BENEFIT-IN-KIND (P11D) REGIME

DEFINITION OF A COMPANY CAR

A company car is defined by s.168(5)(a) of the Income and Corporation Taxes Act 1988 (ICTA 1988) as "any mechanically propelled road vehicle". This definition does not include:

(a) vehicles for carrying loads
(b) vehicles which are not commonly used and accountable as a private vehicle or
(c) motorcycles or invalid carriages.

(A car provided by an employer for an employee's use is usually referred to as a "company car" whether or not it is provided by an incorporated body.)

DEFINITION OF WHAT CONSTITUTES A BENEFIT

If there is no private use of the car, no "benefit" results and the employee would not therefore be taxed as there would be no "imputed" benefit. However, it is common practice to allow employees to use a company car allocated to them for private use as well as for business purposes. It is because of this benefit that company cars are treated as part of taxable remuneration. As such, employees are liable to income tax on this benefit if they are directors, or employees paid more than £8,500 pa (including benefits). The basis of tax liability is not the actual value of the benefit received but an artificial scale of imputed benefits which have, in recent years, been less than the actual benefit. It has therefore been tax advantageous to provide a company car rather than provide the extra taxable income which would be required to enable an employee to provide himself or herself with the same car. The provision of a company car for higher paid executives is now common practice and the majority of new cars in the UK have for some years been purchased as company cars.

It should be noted that car benefit is included in taxable emoluments, which may have the effect of taking lower paid employees into the taxable £8,500+ pa category.

For example, Smith's salary is £7,500 pa and his employer decides to give him a new company car of 1250 cc for business and private use. The taxable value of this car is £2,140 pa. Smith's emoluments therefore become £9,640 pa, which put him into the taxable category for benefit in kind purposes.

THE SCALE RATE TAX CHARGE

The favourable tax treatment of company cars in the past accounts for their widespread inclusion in remuneration packages.

The table below shows the taxable benefit of cars for the tax year 1992/93. The basis on which the taxable benefit is assessed varies according to:

(a) the age of the car; if the car is more than four years old, the taxable benefit is less
(b) the original market value of the car; if the original market value was less than £19,250, the taxable benefit is based on the engine cylinder capacity (see ranges below), otherwise the taxable benefit is based on two bands:
   (i) original market value £19,250–£29,000
   (ii) original market value more than £29,000.

The cylinder capacity of the engine is also an important factor and tax is applied in three ranges:

(a) 1400 cc or less

(b) 1401–2000 cc
(c) more than 2000 cc.

Scale Rates for Taxation of Company Cars 1992/93

Cars with an original market value of up to £19,250 and having a cylinder capacity.

Cylinder capacity of relevant car in cubic centimetres	Age of car at end of year of assessment	
	Under four years	Four years or more
1400 or less	£2,140	£1,460
More than 1400 but not more than 2000	£2,770	£1,880
More than 2000	£4,440	£2,980

Cars with an original market value of more than £19,250.

Original market value of car	Age of car at end of relevant year of assessment	
	Under four years	Four years or more
More than £19,250	£5,750	£3,870
More than £29,000	£9,300	£6,170

The taxable benefits are also modified according to the number of business miles travelled.

1. The taxable benefit is one and a half times the scale charge if the annual business miles travelled are 2,500 miles or less. This increased scale charge applies also to second cars, whatever the business mileage.
2. The car benefit is reduced to half the scale where the annual business mileage is 18,000 or more.

For cars not having a cylinder capacity, the taxable benefits of a car and fuel are both assessed by reference to original market value.

Cars with an original market value of up to £19,250 and not having a cylinder capacity (eg an electric car).

Original market value of car	Age of car at end of relevant year of assessment	
	Under four years	Four years or more
Less than £6,000	£2,140	£1,460
£6,000 or more but not more than £8,500	£2,770	£1,880
£8,500 or more but not more than £19,250	£4,440	£2,980

Definition of "Original Market Value"

The phrase "original market value" means the notional price that would have been paid for the car by a buyer in the UK retail market. Account may be taken of any probable discounts available on a single but not on a multiple car purchase. This notional value is to be determined as at the time immediately before the date upon which the car was first registered.

An effect of this interpretation is that employees do not obtain any reduction in the scale rate tax charge if they are provided with cheaper secondhand cars which are one to three years old.

What is Included in the Scale Rate Tax Charge

The scale rate tax charge includes the initial purchase price of the car and all the usual running costs such as:

- insurance
- road fund licence
- repairs and maintenance
- breakdown cover.

The scale rates do not include items such as:

- accessories fitted to the car after the date of purchase
- mobile telephones
- private car fuel
- chauffeurs for private use.

Time Apportionment of the Scale Rates

Where an employee has a company car available for his or her private use for only part of a tax year (6 April to 5 April following), the scale rate charges are reduced on a time apportioned basis. For example, if the car is available only as from 6 November in any year, the scale rate charge for that tax year will be reduced to 5/12 of the normal scale rate to reflect the amount of time in that tax year during which the car was available for private use.

DIFFERENCE BETWEEN COMPANY CARS AND POOL CARS

A clear distinction must be made between a company car and a "pool" car, as the income tax treatment for the employee is quite different.

Company cars which are held in a "car pool" and are made available for the business use of certain directors and/or employees will not be regarded as having been made available for the private use of any of those directors or employees and will not therefore lead to a personal benefit in kind tax liability provided that the following conditions are met.

1. Any pooled car must be made available to, and actually used by, more than one of the persons in the group to whom it has been made available.
2. The pooled car must not in any year of assessment have ordinarily been used by one of the directors or employees to the exclusion of the others.
3. In the case of each director or employee to whom the pooled car has been made available, any private use of the car must be merely incidental to his or her other use of it.
4. The pooled car must not "normally" be kept overnight on or in the vicinity of any residential premises of any of the directors or employees to whom the pooled car is made available, except that the car can be kept overnight on company premises. (See s.159 of the Income and Corporation Taxes Act 1988 (ICTA 1988).)

Chauffeurs

Where a director or employee paid over £8,500 pa has a company chauffeur made available for his or her private use, this is a taxable benefit in kind. (See s.154 of ICTA 1988.)

Although most items provided in connection with a company car are included in the scale charge, the provision of a driver is expressly outside the scale charge. (See s.155(1) of ICTA 1988.)

The cash equivalent value of a company chauffeur on which the benefit in kind assessment will be made will be the amount of any expense incurred by the company in providing that chauffeur. (See s.156 of ICTA 1988.)

Where chauffeurs are employed to drive "pooled cars" and they are obliged to take a pooled car to their home for overnight parking, the element of travel between their normal place of work and their home will not be regarded by the Inland Revenue as private use and will not thereby disqualify the car from treatment as a pooled car. In addition, overnight parking of the car at the chauffeur's home will not normally be regarded by the Inland Revenue as disqualifying the car from ranking as a pooled car.

Where a director or higher paid employee makes some private use of a pooled car and chauffeur but this use is merely incidental to its business use, that employee will not face a benefit-in-kind charge. (See s.159(2)(b) of ICTA 1988.) However, journeys between a private home and the normal place of work do not rank as business use.

OTHER CAR RELATED BENEFITS

Second and Additional Cars

Where a second or further car is supplied by an employer to an employee, the scale rate applicable is that where less than 2,500 miles are travelled on business in a year.

A director or employee earning £8,500 pa or more will not be taxed on the benefit of a car made available for private use to a member of his or her family or household, provided that the person to whom the car is made available is chargeable on the benefit in his or her own right.

Company Car Parking Spaces

The provision of an allotted car parking space at or near the principal place of work of a director or employee is a valuable benefit, particularly where the offices concerned are in a city centre. The Inland Revenue considered whether the provision of a car parking space should be regarded as a separate benefit in kind and taxed as such but announced in February 1989 that they would abandon this approach.

Car Telephones

Until March 1991, the Inland Revenue took the view that the benefit of a car telephone provided by an employer in a company car was included in the scale charge. The treatment of car telephones provided by an employer for use in a private car was covered in Inland Revenue Statement of Practice SP5/88.

Under the Finance Act 1991 each mobile telephone provided by an employer is to be assessed as a benefit in kind subject to a standard charge. This charge is set at £200. There is no charge if there is no private use of the telephone by the employee or if the employee is required to and does make good the whole cost of any private use, including the appropriate proportion of subscriber charges but not other standing costs.

PERSONAL CONTRIBUTIONS

Some employers allow eligible staff to add a personal contribution to the cost of a car, thereby enabling them to obtain a better model of car or have additional extras. Some important points to consider are set out below.

1. Whether there should be an upper limit to the amount of personal contribution. (Usually this is limited to 25% of the employer's contribution.)
2. If it is a capital contribution, how the employee's sale proceeds from the disposal of the car should be determined.
3. If it is a capital contribution, will this bring the value of the car into the tax band applying to higher value cars, ie more than £19,250 in 1992/93? (It should be noted that the higher rates of tax will apply,

even though the employee has made the contribution himself or herself.)
4. What happens if the employee resigns from the company before the car is due for replacement? If the company does not wish to dispose of the car, there should be a procedure to allow the employee to get back the value of his or her equity in it.
5. Should there be rules to prevent employees from making personal contributions which would enable them to acquire a car appropriate to a more senior grade?

Employees making a contribution towards the running costs, eg by contributing towards the contract hire cost, should be charged VAT on their contributions. The total of their contributions can, however, be set against the scale charge for the taxable benefit of the car.

For example, Smith is allowed a car which costs up to £300 per month in contract hire costs. He chooses a car which costs £320 per month and will pay the extra as a personal contribution. He will be charged £20 per month plus VAT, ie £23.50 per month or £282 in a year. If the scale charge for his car is, say, £2,770, this will be reduced to £2,488 (£2,770 − £282).

COMPANY CARS – THE PROPOSED NEW P11D REGIME

The present system of setting the benefit-in-kind value of the private use of a company car by reference to fixed scale rate charges was first introduced in the Finance Act 1976. Previously the benefit-in-kind level had been determined by the Inland Revenue apportioning the actual cost to the company of providing the car between business and private mileage. This was a time-consuming exercise for both taxpayer and Inland Revenue. The system of scale rate charges was originally designed to be clear, simple and light on administration. It was recognised that there were, inherent in the system, a number of inconsistencies and inequities but, as the initial scale rate charges were modest, the system could bear those flaws as a modest price to pay in exchange for its simplicity and ease of administration.

However, as the present Government steadily and substantially increased the scale rate charges and as car manufacturers began to design car specifications to exploit the weaknesses in the system at the scale rate break points, the flaws became more widely exposed and less easy to tolerate.

The Chancellor of the Exchequer announced at the time of his Budget Statement in March 1992 that he had instructed the Inland Revenue to conduct a detailed review of the system for taxing the benefit of the private use of a company car and that a consultative document would be published in summer 1992.

This consultative document entitled "Company Cars — Reform of Income Tax Treatment", duly appeared at the end of July 1992. In his Foreword to the document, the Chancellor alluded to four facts which offer an interesting pointer to his views:

1. Since the initial legislation was introduced in the Finance Act 1976, the number of company cars has quadrupled to two million cars.
2. The current fixed scale benefit-in-kind charges make a significant £1.4 billion contribution to the Exchequer.
3. The Government's policy on the taxation of company cars is one of broad neutrality so that the tax regime should neither encourage nor discourage company cars nor favour certain car models over others.
4. The Government recognises that company cars are an important feature of modern business life and of the UK car industry (a coded message that company cars are a special interest group).

The consultative document substantially reflects the four pointers indicated in the Chancellor's Foreword. Although the document asked for comment and views from readers to be received by November 1992, it is clear that the eight threads set out below are likely to form the basis of the new tax regime for company cars.

1. Car Price to be Main Determinant of Level of Tax Charge

The current tax bands based on engine size would be replaced by a range of price based bands. These price based bands would use manufacturers' retail list prices (including car tax and VAT), rather than actual purchase price or notional original market value as used at present.

The Inland Revenue believes that a new system based on manufacturers' list prices would be fairer than the present system. The examples are given of a Ford Fiesta 1.8 diesel and a Vauxhall Astra 1.7 diesel (both costing under £9,000) currently falling into the same car scale band as a Mercedes Benz 190 and a BMW 520i (both costing around £19,000).

Four advantages of a new list price based system are identified:

(i) it would not (unlike the present system) discriminate against diesel engined cars which tend to have a higher initial purchase cost than petrol engines of equivalent size
(ii) it would encourage car manufacturers to keep their prices down
(iii) annual adjustments to the car scale rates would no longer be required because the new system would be largely self-adjusting
(iv) the separate scale rates for cars with non-reciprocating engines (eg "Wankel" engines) could be abolished.

2. Tax Charge Largely Determined by a Series of List Price Bands

The Inland Revenue has recognised that to measure the tax charge by reference to every car's list price would lead to a multiplicity of individual calculations. Consequently, to reduce administration, a system of list price bands is proposed. The Inland Revenue's preference appears to be for a series of progressive bands, where each band covers a range of list prices one and one quarter times greater than the immediately preceding band. This approach appeals to the Inland Revenue because it is relatively sensitive in the list price areas in which the large majority of company cars lie. For example, in the table below which is put forward by the Inland Revenue, there are four bands between £8,000 and £19,499, which is the price range for most cars falling into the present single band for cars having engine sizes between 1400cc and 2000cc:

Car List Price (£)		
Up to 4,999		
5,000	–	6,499
6,500	–	7,999
8,000	–	9,999
10,000	–	12,499
12,500	–	15,499
15,500	–	19,499
19,500	–	24,499
24,500	–	30,999
31,000	–	38,999
39,000	–	49,999
50,000	–	62,500

3. Relationship Between the Tax Charge and List Price

Two criteria are proposed for measuring the level of taxable benefit. First, the benefit-in-kind scale rate tax charges would be determined by taking a proportion of list prices. Second, the current business miles "break points" would be retained as would the principle that the "standard" tax charge should be based on business mileage between 2,500 and 18,000 miles with a 50% uplift for lower business mileage and a matching 50% reduction for higher business mileage. The new scale rates proposed by the Inland Revenue are as follows:

Business Use	Business Miles	Scale Tax Charge
Low	Under 2,500	1/3 List Price
Medium	2,500–18,000	2/9 List Price
High	Over 18,000	1/9 List Price

4. Continue Reduced Scale Rates for Cars over Four Years Old

The Inland Revenue proposes to retain the principle of a reduction of around one-third in the scale rate tax charge for cars over four years old. However, the Inland Revenue has become aware of the current "loophole" for classic cars. Under the present regime, account is taken only of the notional original market value of cars and certain classic cars which have substantially appreciated in terms of present market value are taxed at the lowest rates. This loophole will be closed by revaluing cars over 10 years old at the higher of their open and original market values.

5. No Change to Car Fuel Scale Charges

The Inland Revenue believes that engine size is an appropriate yardstick for quantifying the level of benefit of petrol and diesel fuel provided for private mileage and proposes no change in the current system.

6. Monthly Reporting Requirement on Companies

One of the main shocks in the consultative document is the suggestion from the Inland Revenue that employers should be required to make monthly returns of cars newly provided to employees. This is on the pretext that employees do not like additional year-end tax assessments and prefer to pay all tax due through the PAYE system in which case their tax codes will need adjustment in the course of the tax year. In fact this proposal has more to do with increased cash flow for the Exchequer and is in line with the new system of "K" codes which come into effect for PAYE purposes from 6 April 1993 and which enable larger PAYE deductions to be made to reflect the full value of benefits-in-kind, unlike the present system under which the coding deductions for benefits cannot exceed the value of the taxpayer's personal allowances.

7. No Reduction in the £1.4 Billion Tax Raised from Company Cars

There are many references in the consultative document which indicate that any changes introduced by the new regime must not cause a reduction in the current £1.4 billion of tax raised for the Exchequer by the present benefit-in-kind system.

8. Timing of the Change

The Inland Revenue proposes that the new regime will be introduced in the Finance Act 1993 to take effect from 6 April 1994.

WINNERS AND LOSERS UNDER THE PROPOSED NEW REGIME

The table opposite sets out some of the cars which will face either a significant decrease or increase in their associated benefit-in-kind scale rate tax charges. It is not surprising that the biggest reductions will be for the smaller cars and the biggest increases for luxury cars and "perk" cars with engines just under 2,000cc and priced at just under £19,250.

Proposed New Tax Regime — Winners and Losers

Based on 2,500–18,000 Business Miles

Cars	Old Scale Rate £	New Scale Rate £	+/- %
Rover Mini 1.3 Sprite	2,140	1,137	- 47
Ford Fiesta 1.1 CFi	2,140	1,443	- 33
VW Polo G40 1.3	2,140	2,324	+ 9
Ford Escort 1.8 Diesel	2,770	2,030	- 27
Rover 216 SLi	2,770	2,575	- 7
Cavalier 1.6 LS Hatchback	2,770	2,813	+ 2
BMW 518i SE	2,770	4,239	+ 53
Volvo 940 GLE 2.0 Turbo Estate	2,770	4,259	+ 54
Rover 825 Diesel	4,440	3,946	- 11
BMW 520i	5,750	4,377	- 24
Ford Granada 2.0i LX Estate	5,750	4,457	- 22
Range Rover 3.9 Vogue Auto	5,750	6,186	+ 8
Jaguar XJ6 4.0 Sovereign	9,300	8,175	- 12
Porsche 911 Carrera 4 Coupe	9,300	11,808	+ 27
BMW 750iL	9,300	13,833	+ 49

COMPANY CARS – PRIVATE CAR FUEL AND MILEAGE ALLOWANCES

PRIVATE CAR FUEL

Where an employer provides an employee with car fuel for private use, the employee faces an income tax benefit-in-kind charge based on fixed scale rates and the employer incurs a liability to pay national insurance contributions based on the same scale rates (see *Company Cars – The National Insurance Aspects* later in this chapter). The income tax scale rates and concomitant NIC charges which apply for the tax year 1992/93 are set out below:

Annual Business Mileage

Engine Capacity (cc)	Below 18,000 Scale rate (£)	NIC (£)	18,000 miles and above Scale Rate (£)	NIC (£)
Petrol engined cars				
Up to 1400	500	52.00	250	26.00
1401–2000	630	65.52	315	32.76
Over 2000	940	97.76	470	48.88
Diesel engined cars				
0–2000	460	47.84	230	23.92
2001+	590	61.36	295	30.68

Cost Effectiveness of Private Car Fuel

The tables opposite show the actual benefit compared with the taxable benefit of the employer paying the petrol or diesel costs of the employee. Actual benefits are based on the AA's figures and assume that petrol is £2.27 per gallon and that diesel is £2.05 per gallon, that 10,000 miles are travelled pa and that business mileage is more than 2,500 miles pa. Figures relate to 1992/93. Once again the effect of tax has to be taken into account in assessing the value of the employer's contribution. The tables show the effect for these in 40% and 25% marginal tax rates; the effect of the employer's NIC cancels out in the ratio.

What is shown in the last column is the break-even mileage, the mileage at which the taxable benefit equals the actual benefit. Fuel benefit becomes tax efficient, given the assumptions above, only when the annual mileage exceeds that shown in the break-even column.

The break-even mileage decreases as the price of petrol increases. Bearing in mind the increasing costs of petrol, payment by the employer of petrol costs for private motoring may have become more attractive of late. Unleaded petrol, being cheaper than leaded, means that in general break-even mileages will be higher than for cars running on unleaded petrol.

COMPANY CAR MILEAGE ALLOWANCES

Where directors or higher paid employees have use of company cars but do not receive all their petrol paid by their employers, they are entitled to

Taxation of Company Cars

Petrol Benefit

Marginal tax rate 40%

Motor vehicle engine size (cc)	Petrol cost (£)	Pre-tax value (£)	Taxable value (£)	Pre-tax value divided by taxable value	Break-even mileage pa
1000	567	945	500	1.89	5,291
1001–1400	649	1,080	500	2.16	4,629
1401–2000	756	1,260	630	2.00	5,000
2001–3000	1,031	1,720	940	1.83	5,464
3001–4500	1,135	1,891	940	2.01	4,975

Marginal tax rate 25%

Motor vehicle engine size (cc)	Petrol cost (£)	Pre-tax value (£)	Taxable value (£)	Pre-tax value divided by taxable value	Break-even mileage pa
1000	567	756	500	1.51	6,622
1001–1400	649	864	500	1.73	5,780
1401–2000	756	1,009	630	1.60	6,250
2001–3000	1,031	1,375	940	1.46	6,849
3001–4500	1,135	1,513	940	1.61	6,231

Diesel Benefit

Marginal tax rate 40%

Motor vehicle engine size (cc)	Fuel cost (£)	Pre-tax value (£)	Taxable value (£)	Pre-tax value divided by taxable value	Break-even mileage pa
0–2000	455	758	460	1.65	6,060
2001 + (A)	586	977	590	1.66	6,024
2001 + (B)	683	1,138	590	1.92	5,208

Marginal tax rate 25%

0–2000	455	607	460	1.32	7,575
2001 + (A)	586	781	590	1.32	7,575
2001 + (B)	683	911	590	1.54	6,493

Note A: cars with a new purchase price not exceeding £15,000
B: cars with a new purchase price exceeding £15,000

claim the actual costs of petrol and oil incurred on business journeys as a business expense.

Whether the petrol and oil consumed on business journeys are purchased through a company credit card or paid for privately and then reclaimed through the employer's expenses system, the expenditure should be justified separately, preferably on an expenses claim form. It may be easier for the employer to publish a fixed mileage allowance based on statistics published from time to time by the major motoring organisations, varied to reflect current petrol and oil costs. An allowance, at the date of publication, might be given as follows:

Motor vehicle engine size	Allowance per mile
Up to 1000 cc	6p
1001 to 1400 cc	7p
1401–2000 cc	8p
2001–3000 cc	10.5p
3001 cc and above	12p

(These figures are based on a petrol price of £2.27 per gallon. Employers should adjust allowances in line with alterations in petrol prices.)

PRIVATE CAR MILEAGE ALLOWANCES

Where directors or employees use their personal cars on company business, their employers will expect to pay them a car mileage allowance. Provided that this allowance does not contain any profit element, the Inland Revenue will accept that such payments are tax-free.

In June 1990, the Inland Revenue announced the introduction of a "Fixed Profit Car Scheme" (FPCS) to simplify the existing arrangements and to reduce the amount of work required of employers, employees and the Inland Revenue.

The intention of the law on private car mileage allowances is that any profit employees make on the excess of an employer's mileage allowance over actual costs should be taxable. Strictly, the allowances should be reported by the employer on either Form P11D or P9D, and the employee should send a detailed expenses claim to show the total motoring costs incurred. Tax relief is then given on the proportion of tax allowable costs related to business mileage.

Previously, the Inland Revenue informally indicated the mileage rates which it accepted as containing no profit element. These were related to the engine size of the car. It has now formalised this arrangement and made a new provision for employees who use their car over 4000 miles pa for business purposes in the Fixed Profit Car Scheme. For 1992/93 the FPCS tax-free rates are:

Motor vehicle engine size	Up to 4,000 miles	Over 4,000 miles
up to 1000 cc	25p	14p
1001-1500 cc	30p	17p
1501-2000 cc	38p	21p
Over 2000 cc	51p	27p

It should be emphasised that these allowances apply only to business use of a private car; travel from home to work and other private mileage should be excluded. Dispensations from reporting are available from tax offices if mileage allowances are less than those in the table above. Where employers pay the same rate of mileage allowances whatever the size of car the employees use for business, the average of the two middle bands is used. In 1992/93 this is 34p for the first 4,000 miles and 19p thereafter.

FPCS is voluntary; employers do not have to use it and even where they do, any employee may choose to have his or her mileage allowance taxed on the strict statutory basis. This would clearly be appropriate if total running costs of the private car were greater than those allowed in the Inland Revenue's tax-free mileage rates.

COMPANY CARS – WHAT IS A BUSINESS MILE?

The income tax, national insurance and value added tax regimes as they apply to company cars are based upon the ability of companies and the car drivers themselves to identify accurately what is, and what is not, business travel. In fact, because the law on the point is somewhat abstruse, the distinction is not that simple to make. On the surface, the distinction to be made is clear-cut and lies between business and private mileage. This is too simple a distinction and leads to most of the disputes that arise. On the one hand the fiscal authorities argue that at law certain car usage cannot be regarded as business travel and must therefore be regarded as private mileage. On the other hand the employee argues that the car journey at issue was only made in furtherance of his or her duties as an employee and, as there was no personal benefit from the journey, it is unfair to construe it as personal mileage. The position becomes clearer once it is appreciated that there are in fact three rather than two categories into which car mileage can fall. These are:

(a) journeys which both employees and the law regard as business travel
(b) journeys which, because they are either directly related to the duties of their employments or undertaken to enable the better performance of those duties, employees (and perhaps common sense would also) regard as business travel but which the law does not accept as business travel
(c) journeys which both employees and the law regard as private mileage.

Only category 1 counts as business travel for income tax, national insurance and VAT purposes. As an added complication, there will inevitably from time to time be mixed motive journeys. These must be fitted into one of the three above mentioned categories. There is set out below a brief interpretation of the law relating to business travel and how it applies to categorising car mileage.

Definition of "Business Travel"

Business travel is defined for the purposes of the company car benefit-in-kind legislation as "travelling which a person is necessarily obliged to do in the performance of the duties of his employment" (see s.168(5) (c) of the Income and Corporation Taxes Act 1988 (ICTA 1988). This mirrors the provisions allowing directors and employees to deduct, *inter alia*, business travel expenses from their taxable incomes. These provisions state that "If a holder of an office or employment is necessarily obliged to incur and defray out of the emoluments of that office or employment the expenses of travelling in the performance of the duties of the office or employment . . . there may be deducted from the emoluments to be assessed the expenses so necessarily incurred and defrayed" (see s.198(1) of ICTA 1988). This is helpful because the past judicial interpretations of s.198(1) can be used to determine how the courts would be likely to interpret s.168(5)(c).

There are two tests to satisfy if car mileage is to qualify as business travel. The employee must be both (i) "necessarily obliged" to undertake the journey and (ii) travelling "in the performance of his employment".

"Necessarily obliged"

The Court of Appeal has determined that the first test means not that a specific employer requires the employee to make the journey nor that the employee makes the journey believing it is necessary for the better performance of his or her duties. The test is that the duties of the employment require him or her to make the journey. In other words, the test is that any employee who fulfilled the duties of the particular employment would necessarily be obliged to undertake the journey. This interpretation might be used for example to disqualify a long journey to a remote company location undertaken towards the end of a tax year which just helps the employee to pass the 2,500 business miles hurdle.

"In the performance of his employment"

The second test is that whilst the journey is undertaken, the employee must be performing the duties of his or her employment. It is this hurdle which is used to disallow travel between home and the place of work. This is a classic example of category 2 above in that the employee only undertakes the journey because he or she needs to get to and from his or her place of work and there is no personal benefit as such but, nonetheless, it does not constitute business travel. The general rule is that the duties of an office or employment do not commence until arrival at the place of work and that therefore car mileage prior to arrival at the place of work cannot be business travel. This test can be harsh and has even disqualified car mileage to hospitals of doctors responding to emergency calls.

Mixed motive journeys

Fortunately, unlike other business expenses, a journey does not have to be undertaken "wholly and exclusively" in the performance of the duties of the employment so mixed motive trips can still qualify as business travel. If a journey had to be made by an employee based normally in

Luton on a Friday to, say, the company's plant in Edinburgh, it could still properly be categorised as business travel even if the employee chose to stay in Edinburgh for the weekend for personal sightseeing purposes.

Only personal business travel counts towards mileage thresholds. Where an employee has a company car specifically allocated for his or her use, only the business travel undertaken personally counts towards the 2,500 and 18,000 business mile thresholds. Loaning the car to other employees for their business travel does not "earn" the business miles driven by those other employees (see paragraphs 3(1) and 5(1) of Schedule 6 ICTA 1988).

Two or More Company Cars

Where an employee has more than one company car only that car which has been used for the most business travel can pass the 2,500 or 18,000 business miles threshold. Regardless of the business mileage actually driven in them in the course of any tax year, the second and subsequent cars will be treated as having been driven for less than 2,500 business miles and the benefit-in-kind scale rates will be increased by 50%.

COMPANY CARS – THE NATIONAL INSURANCE ASPECTS

Employers are, from the tax year commencing 6 April 1991, required to pay national insurance contributions (NIC) on the benefit of company cars available for private use. The basis of liability is assessed on a similar basis to income tax, in that:

- no contributions are levied unless an employee's emoluments including benefits exceed £8,500 pa
- the NIC liability is determined by using the scale rates published by the Inland Revenue.

Contributions are levied at the main rate of 10.4% in 1992/93 and are assessed annually. The payments for 1992/93 will be due in June 1993.

Employers' NIC payments on company cars under four years old in 1992/93 will be:

Annual Business Mileage

	Below 2,500 miles £	2,500–18,000 miles £	18,000 miles and above £
Engine capacity (cc)			
Up to 1400	333.84	222.56	111.28
1401–2000	432.12	288.08	144.04
Over 2000	692.64	461.76	230.88
£19,250–29,000	897.00	598.00	299.00
Over £29,000	1,450.80	967.20	483.60

The table below shows the NIC payments for cars which are more than four years old.

Annual Business Mileage

	Below 2,500 miles £	2,500–18,000 miles £	18,000 miles and above £
Engine capacity (cc)			
Up to 1400	227.76	151.84	75.92
1401–2000	293.28	195.52	97.76
Over 2000	464.88	309.92	154.96
£19,250–29,000	603.72	402.48	201.24
Over £29,000	962.52	641.68	320.84

Employers' NIC liability also extends to the benefit of free fuel provided for private use in company cars. Again, Inland Revenue scale rates are used as the basis for the NIC charge.

The Finance (No. 2) Act 1992 introduced a separate lower scale rate for diesel fuel. The scale rates and NIC payments are as follows for 1992/93:

Annual Business Mileage

	Below 18,000 miles		18,000 miles and above	
Engine capacity (cc)	Scale rate £	NIC £	Scale rate £	NIC £
Petrol engined cars				
Up to 1400	500	52.00	250	26.00
1401–2000	630	65.52	315	32.76
Over 2000	940	97.76	470	48.88

Annual Business Mileage

Engine capacity (cc)	Below 18,000 miles		18,000 miles and above	
	Scale rate £	NIC £	Scale rate £	NIC £
Diesel engined cars				
0–2000	460	47.84	230	23.92
2001 +	590	61.36	295	30.68

HOW TO CALCULATE THE NIC LIABILITY

As stated above, the burden of NIC in respect of cars and car fuel falls solely on the employer. There are four principal factors to take into account when calculating the amount of NIC due:

(a) the income tax scale rates
(b) the amount of business mileage
(c) the availability of the car and/or car fuel during the tax year
(d) any cash contributions made by the employee during the tax year.

Income Tax Scale Rates

These have been set out above both for company cars and for car fuel. It should be noted that these scale rates and the NIC liability apply only to cars (including estate cars and electric cars) but not to motorcycles, invalid carriages or commercial vehicles such as lorries, trucks and vans.

The Amount of Business Mileage

The employer is required either to pay NIC at the maximum rate for business mileage under 2500 miles pa or to keep detailed mileage records to support making payments at the lower rates on the basis of higher business mileage.

This is where the weakness of hanging the NIC charge on the back of the Inland Revenue income tax scale rates is exposed. The scale rate system based on business mileage worked for income tax purposes because employers could leave employees to make their own returns of business mileage to the Inland Revenue on their personal annual tax returns and, by completing the "Don't know" box on form P11D (annual return of benefits-in-kind), avoid direct involvement in the determination of business mileage.

However, because the level of an employer's NIC liability in respect of company cars depends upon the amount of annual business mileage, the employer now has to get involved in monitoring the business mileage of all employees who drive company cars. This means determining into which of the three categories of business mileage each company car falls and declaring the category on form P11D, for the "Don't know" box has now disappeared from the form. This is a time consuming and expensive exercise. This situation is not helped by the further difficulties of determining what are, and what are not, business miles (see *Company Cars – What is a Business Mile?* above).

Availability of the Car/Car Fuel

The NIC charge will be reduced if the company car is available for only part of a tax year. The reduction will reflect the proportion of the tax year during which the benefit was not available. So if, for example, an employee does not receive a company car until 6 July, the NIC charge will be reduced by 25% to reflect that it was not available for the first three months of the tax year. A reduction is allowed where a car is not available because of extensive maintenance or repairs but the continuous period for which the vehicle was not available because it was not capable of being used must be 30 or more days. The NIC charge on car fuel is linked to availability of the company car and not directly to the provision of car fuel. This can lead to a full year's charge where car fuel is discontinued as a benefit in the course of a tax year but the company car remains available.

Employees' Cash Contributions

Where an employee makes a cash contribution to his or her company in respect of the private use of the company car, the scale rates are reduced by the amount of the contribution and the NIC liability is calculated on the reduced scale rate. Unfortunately, the NIC liability on the benefit of car fuel cannot be partially reduced – if the employee makes a full reimbursement of his or her private car fuel costs, the NIC liability is entirely extinguished but if only a partial reimbursement is made, no reduction in the NIC charge is allowed.

Collection of NIC Charges

Employers must pay their NIC charges on company cars and car fuel by 19 June in each year in respect of the immediately preceding tax year.

COMPANY CARS — VALUE ADDED TAX

Value added tax ("VAT") is a significant element in the costs of providing company cars and has five areas of relevance:

- buying and selling the car
- financing the car
- running costs
- car fuel
- employee contributions to costs.

These are considered separately below.

Buying and Selling the Car

VAT is charged on the purchase price of the car and is not recoverable by the employer. This VAT liability applies also to delivery charges and the cost of any accessories fitted to the car at the date of purchase.

On the sale of a car, there is no VAT liability on the sale price, except in the rather unlikely event that the sale price exceeds the total initial purchase price including VAT. However, where a profit is realised, VAT is due on the profit element and is calculated in the normal manner which at the current standard rate is 17.5/117.5 of the profit.

Financing the Car

Where a company does not buy the car out of its own cash resources or through a bank loan, VAT will apply to the selected financing route as follows:

(a) *Contract Purchase/Hire Purchase/Lease Purchase.*
Where the company has a right to buy the car at some future date, the situation is very similar to that applying to an outright purchase of the car. VAT is payable on the total purchase price of the car, is not reclaimable by the finance company and is therefore factored into the hire or rental charges. However, VAT is not payable in respect of the interest element of the hire or rental charges. For this reason, these financing routes may be appropriate for employers who

cannot recover VAT because they are wholly or partially exempt.
(b) *Contract Hire/Finance Lease/Temporary Rental.*
Where the company hires or leases a car and does not have a future right to buy the car, VAT is payable on the whole of the hire or rental charges. The input VAT payable on these charges is recoverable but only by those companies which have sufficient output VAT to use as off-set. For this reason, these financing routes are likely to be less attractive to employers who are wholly or partially exempt for VAT purposes.

Again, the finance company cannot recover the VAT payable on the purchase price of the car, so this element of VAT cost will be factored into the hire or rental charges.

Running Costs

Input VAT is recoverable where it is incurred in connection with running costs and accessories fitted to the car after the date of purchase.

Car Fuel

Input VAT is recoverable in respect of car fuel and where employers pay fixed mileage allowances for business mileage (see *Company Cars – Private Car Fuel and Mileage Allowances* above), on the proportion of the allowance which is directly attributable to the cost of the car fuel but not engine oil.

VAT scale charges apply where petrol or diesel fuel is provided to employees for private mileage. These scale charges which are set out in the table overleaf are based on the income tax scale rates for private car fuel and allow for modestly lower charges for diesel.

High business mileage

Under the VAT as well as the income tax and NIC regimes, the scale charges are halved for high business mileage and the annual break point for all three regimes is the same at 18,000 miles. However, whereas for income tax and NIC purposes the test is measured on 18,000 business miles over the course of a tax year, for VAT purposes the test has to be met for each VAT period. Consequently, if the employer files quarterly VAT returns the hurdle becomes at least 4,500 business miles in each quarter and if monthly returns are filed the hurdle is 1,500

Petrol

VAT Return	Car Engine Capacity (cc)	Scale Charge (inc. VAT) £	VAT Element £
Quarterly	Up to 1400	125.00	18.62
	1401-2000	158.00	23.53
	Over 2000	235.00	35.00
Monthly	Up to 1400	42.00	6.26
	1401-2000	53.00	7.89
	Over 2000	78.00	11.62

Diesel

VAT Return	Car Engine Capacity (cc)	Scale Charge (inc. VAT) £	VAT Element £
Quarterly	Up to 2000	115.00	17.13
	Over 2000	148.00	22.04
Monthly	Up to 2000	38.00	5.66
	Over 2000	49.00	7.30

business miles each month. Averaging over the course of a year is not permitted for VAT purposes so that if annual business mileage is, for example, 20,000 miles divided as 8,000 miles in Q1, 4,200 miles in Q2, 3,000 miles in Q3 and 4,800 miles in Q4, the reduced scale rates for an employer making quarterly VAT returns will only be available in Quarters 1 and 4.

Employer responsible for VAT on private car fuel

The burden of the income tax liability for private car fuel based on the scale rates falls on the employee but, as with NIC, the VAT liability falls on the employer. Where private car fuel is provided, the employer is required to keep detailed records of the cars concerned, the employee drivers and of high business mileage where the 50% reduction is claimed.

Where total company mileage is low or the employer is partially exempt and therefore unable to recover input VAT on car fuel in full, it may be advantageous to opt out of the VAT regime on car fuel. If an employer so elects, no input VAT is recoverable on any company purchase of car fuel whether for business or private use but no VAT is payable either on the scale charges.

Employee Contributions to Costs

In certain instances, employees make a personal contribution to the costs of their company cars. In these cases the employer is treated as having made a "VATable" supply and the payments from employees are regarded as including VAT. For example, where an employee contributes £100 per month by way of salary deduction, his or her annual payment of £1,200 is deemed to comprise £1,021.28 contribution to the cost of the car and £178.72 VAT. Where the employee contribution is intended to extinguish any income tax or national insurance contribution liability, account must be taken of this VAT element. The problem is particularly acute as regards private fuel where any reimbursement by the employee of less than the full scale charge amount may be disregarded for national insurance and VAT purposes.

CASH OR CAR SCHEMES

For some time it appeared that H M Customs & Excise were taking the point of view that where employees were offered a choice between a company car or a cash allowance, those who elected for the car would have to pay VAT calculated on the value of the cash alternative. Fortunately, this threat has disappeared with the introduction of new regulations and with effect from 1 April 1992, VAT is not levied on cash or car schemes drafted to fall within the exemption. Care must still be taken to ensure that cash alternatives to car schemes do not fall into the VAT net. Employee cash contributions to the costs of a company car remain subject to VAT and any scheme whereby salary or wages foregone in return for a company car could be interpreted as "back-door" rental payments would be vulnerable.

Chapter 9
BUYING OUT THE COMPANY CAR

When the fixed scale rates for taxing the benefit of the personal use of a company car were first introduced in the Finance Act 1976, they represented a bargain for company car users. This conspicuous bargain value remained for just over a decade and contributed substantially to the rapid spread of the company car as a standard benefit for management, professional and supervisory grade employees. As stated earlier, the number of company cars has increased over the period 1976 to 1991 from 500,000 to 1,900,000 cars.

However, the present Government has embarked upon a policy of tax neutrality and of removing areas of artificial tax advantage. The removal of the tax advantage for company cars began in the Finance Act 1988 when the scale rates were doubled. This trend continued in successive Finance Acts until the Finance Act 1991 whereupon, except for the more expensive cars which are tackled in the new proposals for taxing company cars (see *Company Cars – the Proposed New P11D Regime* earlier), fiscal neutrality appeared to have been achieved. The extent of the increases can be appreciated from the comparison below between the present standard scale rates and those applying in the tax year 1987/88:

Car Engine Size	Tax Year 87/88 £	Tax Year 92/93 £
Under 1400 cc	525	2,140
1401 to 2000 cc	700	2,770
Over 2000 cc	1,100	4,400
Expensive Cars		
Over £19,250	1,450	5,750
Over £29,000	2,300	9,300

After such an extended period of tax advantage, it is not surprising that it should take some time for the appreciation by employees that the "perk" — as opposed to the business "need" — car, is no longer necessarily a significant financial benefit. In the aftermath to the Finance Act 1991, a number of employers began to offer cash allowances in lieu of a company car but found there were few takers of the cash. There were a range of reasons for this. First, employees had not then fully grasped that the substantial tax advantage era had gone. Second, employees like the freedom of having no personal responsibility for their cars. Third, the cash allowances were perceived as being ungenerous and inadequate to cover the net of tax costs of running their own car.

A further "hiccup" for cash allowances in lieu of company car schemes was delivered by H M Customs & Excise which argued for a while that, in certain circumstances, such schemes could trigger a VAT liability on those employees who elected for the car. Fortunately, this threat has now been removed (see *Company Cars – Value Added Tax* in Chapter 8), but it served to delay the introduction of more cash based car allowance schemes.

CAR ALLOWANCES

It is axiomatic that cash allowances in lieu of a company car will not be taken up if voluntary, or will be unpopular if imposed, unless the allowances paid are perceived by employees as covering the costs of running their own car. The problem is that it is quite difficult to calculate allowances that are fair to both employer and employees and absolutely impossible to devise a standard cash allowance programme that is equally fair to all employees whatever their cars, their personal mileages and their individual tax rates.

However, a reasonable starting point is to calculate the cash allowances that the employer could offer on the basis that these allowances should cost neither more nor less than the current cost of providing a company car. Once the size of allowances has been calculated on this basis, they can be measured in terms of whether or not they would constitute adequate compensation for the loss of the company car.

Step 1 — Establishing Base Assumptions

The first essential step is to lay out the constant assumptions to be used throughout the exercise. A set of model assumptions is laid out below:

Buying Out the Company Car

A:	Car Model	Ford Sierra
B:	Engine Size	1800 cc
C:	Initial Cost	£11,000
D:	Retention Period	three years
E:	Resale Value	£5,000
F:	Annual Mileage	10,000 miles
G:	Business Mileage	3,000 miles
H:	Running Costs (servicing/tyres, etc — no petrol) F × 8.84p per mile	£884
I:	Annual Depreciation C - E/D	£2,000
J:	Financing Costs	10%
K:	Corporation Tax	33%
L:	Employers' National Insurance	10.4%
M:	Income Tax - basic rate	25%
	- higher rate	40%
N:	Car Scale Rate — Standard Benefit	£2,770
O:	Road Fund Licence	£110
P:	Insurance	£550
Q:	Business Mileage Allowance (fixed profit car scheme) G × 38p	£1,140
R:	Business Petrol G × 100 gallons at 227p gallon	£227

This completes Step 1. Obviously these assumptions should be updated regularly and varied to fit the circumstances of each employer, each employee and each car.

Step 2 — Calculating the Cost to the Company

The next step is to use the above assumptions to calculate the annual cost to the employer of providing the company car:

Item		£
Finance Cost	(C × J)	1,100
Annual Depreciation	(I)	2,000
Running Costs	(H)	884
Business Petrol	(R)	227
Insurance	(P)	550
Licence	(O)	110
Direct car costs		4,871
Employer's NIC	(N × L)	288
		5,159
Less Corporation Tax relief (£5,159 × 33%)		(1,702)
Net cost to company		3,457

This calculation provides the employer with the cost constraints within which cash allowances can be offered on a no-gain, no-loss basis.

Step 3 — Tax Cost to Employee of Company Car

Of course, the employee receives a benefit-in-kind charge for his or her private use of the company car and this personal cost should be taken into account when measuring the adequacy of the cash allowance.

	Taxpayer Net Cost	
	25%	40%
	£	£
Benefit-in-kind charge (N)	693	1,108

Now it is possible to move to the final stage, the measurement of whether employees will be better or worse off from receiving a cash allowance equal to the present cost to the company of providing the company car.

Step 4 — Measurement of Adequacy of the Cash Allowance

For the purposes of the example below, it is assumed that the employee concerned will not pay national insurance contributions on his or her cash allowance because he or she already pays the maximum personal contributions. It is also assumed that the employee would buy an identical car to his or her present company car and that the cost to him or her of running that car would be the same as for the company on the basis that the employer could put in place group financing and insurance plans.

		Allowances	
		25% **Taxpayer**	40%
		£ **Net Receipt**	£
Cash Allowance	5,159		
Less:			
Mileage Allowance	(1,140)		
Employers' NIC	(379)		
	3,640	2,730	2,184
Mileage Allowance		1,140	1,140
Tax Allowance on Finance Cost (£1,100 × G/F)		83	132
		3, 953	3, 456
Net position of employee			
Direct Car Costs		4,871	4,871
Less:			
Allowances		(3, 953)	(3, 456)
Tax charge avoided		(693)	(1, 108)
Net disadvantage		(225)	(307)

The example above is very sensitive to changes in the underlying assumptions but is a useful starting point and highlights how close the current scale rates are to break-even.

CAR LOAN SCHEMES

Employers are aware that one of the main barriers to establishing cash allowances in lieu of company cars is the practical problem of employees finding the finance to fund the purchase of their own cars. On the face of it, the simplest solution would be for the employer to offer cheap or interest-free loans. There are a number of problems, however, with this route. First, employees may face tax charges in respect of the benefit of their loan. Second, the employer may not itself be able to offer such loans from its own balance sheet. Third, because of Companies Acts restrictions, directors may not be able to receive such loans. Fourth, loans may involve a considerable degree of internal administration especially where they are repaid by salary deductions. Finally, loans may be difficult to collect when employees leave the company.

For these and other reasons, many employers who have considered an alternative loan-based scheme to the provision of a car have arranged an external source of finance. The outside provider of funds may receive some comfort from the employer who may subsidise the interest rate, and the loan facility may depend upon the employee's salary being paid into a bank account held with the lender from which loan repayments can be made direct. The loan/cash programme allowance may be constructed along the following lines.

Year 1

- an initial three year loan through a third party of sufficient size to buy the car; plus
- a cash allowance to cover running costs.

Year 2
A larger cash allowance to cover:

- car running costs
- interest on loan
- loan instalment repayment; and
- income tax on cash allowance.

Year 3
A cash allowance to meet the same items as in Year 2.

Year 4
A back-to-back scheme commences with a lower loan to buy the new car reflecting the residual value of the old car.

Chapter 10
THE FUTURE FOR THE COMPANY CAR

IMPLICATIONS OF GOVERNMENT POLICY

The Inland Revenue has issued a consultative document on company cars to discuss potential reforms of the income tax treatment (see Chapter 8).

The Government has four clearly stated aims:

1. To ensure that, as far as possible, the tax charge people pay is a fair reflection of the benefits received.
2. To avoid tax-driven distortions in the car market.
3. To minimise incentives to drive less fuel-efficient cars.
4. Commensurate with these aims, to keep to a minimum the costs of administering the system, both for employers and the Inland Revenue.

The Government explains its position in some detail but it should be remembered that their overall policy is to approach the taxation of benefits in kind so that there is no difference in the treatment between pay received in cash and pay received in kind. They also feel that the level of tax generally on the provision of cars is now realistic and that there is no incentive either to increase the level of taxation on cars or to decrease it. Change is therefore likely to occur in the way in which tax is collected rather than its amount. In addition to the tax implications, the Government has a stated policy on improving the environment. This, in time, could lead to more incentives to company car drivers and the managers of company cars by making certain choices cheaper to operate. For example, unleaded petrol is cheaper than leaded and diesel is cheaper still. Further encouragement from the Government regarding tax on more fuel-efficient engines is expected along with the possibility that penalties will be applied for those vehicles seen to be environmentally unfriendly.

On the tax implications of road building, there are a number of initiatives being taken by Government at present, concerning the provision of toll roads where, as in most other European countries, faster through-routes are available for those vehicles willing to use them for a premium cash payment. Such charges will have a direct bearing on the cost of running company cars.

The Government further seek to dissuade individuals from driving regularly into the centre of major conurbations and have, for some time, considered imposing restriction zones through which vehicles can only pass at certain times of the day on payment of an additional road fund tax. Such systems have been in operation in a number of overseas countries for many years and have been found to be relatively easy to manage and to be cost-effective.

There is much talk of office relocation away from the centre of conurbations and many individuals have expressed a desire to work from home, which is becoming more practical as a result of the improvements in communication technology. There are many benefits in this for the individual, not least of which is the saving of private mileage use when driving from home to office. It may also make it easier to identify the true business miles run by a vehicle by comparison to the private miles the individual may choose to drive. At present, the Government's policy is to tax the benefit in kind and *based on the business use* this is unlikely to change in the present review. However, perhaps it would be more reasonable to tax the individual on the *private use* of any given vehicle as this is truly a benefit in kind.

At present company car drivers who, by the nature of their work, only drive a few private miles each year can be heavily penalised for the low business mileage.

There has been much debate over the last few years on the tax incentives that may be applied to encourage individuals to use public transport but it would seem more likely that the approach will be directed more towards penalising company car drivers for using vehicles where alternative methods of public transport are available rather than providing an incentive to use public transport.

The car industry is a very significant part of the British economy and as it is estimated that over £1.4 billion is collected each year by the Government on the income tax scale charges alone, the Government is unlikely to wish to severely disrupt it. With Japanese car production plants being added to our existing UK vehicle manufacturers, the benefit to the whole economy from the supply of component parts and the assembly of vehicles, let alone the tax implications, cannot be over-emphasised especially as modern technology in Britain is now enabling the industry to export considerable quantities of cars abroad.

It is therefore likely that the Government will proceed with great caution to prevent unnecessary upheaval in an important industry where the manufacture and provision of new cars has such an impact on the overall economy and where 63% of the production of new cars is sold to companies for use by company car drivers.

THE "REAL" CHALLENGE

The company car fleet management business is much more complex than most companies believe. The benefit of getting it right can mean:

- considerable savings to the company
- a strong company image
- good cash flow management
- best whole-life cost and reliability.

In order to keep ahead of company demands, it is advisable to carry out a company car fleet audit. Whilst it could be recommended that each aspect of the previous chapters be investigated, certain key issues will be particularly relevant to a company and the following fleet audit checklist may help to focus your attention.

COMPANY CAR FLEET AUDIT

At regular intervals take stock of the fleet policy within the company to try and highlight areas that require closer inspection for greater efficiency and lower cost.

The main objectives of the audit could be summarised as follows.

(a) to undertake a review of the need, acquisition, operation and disposal of the fleet vehicles
(b) to identify and quantify areas where costs could be saved

(c) to review the overall strategic fleet policy to determine whether it still meets the overall company objectives
(d) to identify the management needs and review the operational criteria necessary for effective control.

The Present Position

An essential part of any audit is to evaluate fully the present status of the fleet, the range of vehicles being operated and the method of acquisition, servicing, disposal and management. The process by which cars are allocated to individuals needs to be clearly stated and the logic of vehicle selection shown.

The replacement policy should be compared to the actual data gathered on cost of operation to determine whether the policy makes practical sense.

Accident Control

This is best viewed from the insurance broker's perspective to establish what premiums or excess premiums are being paid. The accident history record should be reviewed and management action taken to penalise the major culprits. Line management should be made aware of any investigation so that suitable remedial action can be taken against persistent offenders. Company driver training (including the Institute of Advanced Motorists) should be considered as this has been shown to reduce accidents and can enable discounts to be obtained on insurance premiums.

Fuel Management

This is best considered by use of one of the specific fuel credit card systems that limit the use of the card to fuel only. If not already introduced, then fuel management control in this manner alone can save up to 8% of fuel costs and make administration considerably easier.

Fleet Disposal

Fleet disposal routes (of vehicles owned by the company) need to be reviewed. The most readily auditable route is through a specialised and reputable auction company who will provide a comprehensive valeting service to ensure maximum residual values are achieved and agree a reserve price.

This can be a precise way of offering ex-fleet vehicles to employees as they can go to the auction and bid for the vehicle. Although not necessarily the most profitable means of disposal for the company, it does eliminate many problems such as honesty and favouritism that can otherwise occur. Naturally, for vehicles on contract hire, the problem does not arise as the vehicles are simply returned at the end of the agreed term.

Strategic Economic Issues

These could include:

(a) the provision of secondhand vehicles, perhaps on contract hire, to reduce depreciation costs
(b) down-sizing of vehicles to reduce company operating costs and, at the same time, reduce the tax liability for the users
(c) a review of the implications of "whole-life" costs rather than the initial purchase price of the vehicle
(d) check that the policy to introduce new models into a fleet range does not encourage higher standards of specification to be set within the company car allocation structure, ie ensure that down-sizing of engine capacity is not negated by up-specification of the interior
(e) undertake a review of the car v cash options and ensure that all job descriptions and terms and conditions of employment are written in such a way as to ensure compliance with the law and an ability to introduce change where necessary.

These are only a few of the *key* issues.

There is no short cut to good fleet management control as each company has its own position to consider on every aspect from acquisition to disposal. Although the financial implications will be seen, by the company, to be the most significant feature, the status (or lack of it) perceived by members of staff given cars that do not match up to their expectations, will generate the most aggravation for the fleet manager. (The "psychology of the company car driver" will, however, have to wait for a further book to be dedicated to the subject!)

APPENDIX

SPECIMEN EMPLOYEE BOOKLET FOR THE ABC COMPANY CAR SCHEME

Comments relating to the content of the booklet are shown in square brackets [].
Example of a booklet to be circulated to participating staff.

[It should be noted that the employer is assumed to own and run his or her own car fleet via the personnel department; the appropriate changes would need to be made in this booklet if cars are provided via contract hire or a fleet management service runs the scheme. The terms of the scheme relate to status cars for middle ranking managers; they might be tightened for job requirement cars or relaxed for more senior managers.]

This booklet is for the guidance of staff eligible to receive a company car. The basis of the scheme is that you will be provided with the company car of your choice within the scheme guidelines. The Company will pay or reimburse all running expenses, with the exception of petrol and oil. For your part, you are expected to see that the car is properly used, maintained and cared for.

The car is insured for you to drive. Your spouse (or other nominated drivers over the age of 25) may drive it with your permission. Please contact the personnel department if you wish the cover to be extended to further drivers or have any other queries on insurance, eg green card cover whilst driving abroad.

It is a condition of our insurance that you report all accidents involving your company car to the personnel department, even if no insurance claim is expected.

Please read this booklet carefully, so that you are familiar with the rules and requirements of the car scheme.

1. Car Entitlement

1. Staff in grade 7 and above are eligible to receive a company car. You will be able to choose one from a range of popular models, according to your grade. Staff in more senior grades are allowed greater choice within certain cost limits. Leaflets CAR 1 and CAR 2 at the end of this booklet give current levels of entitlement. [Practical tip: it is a good idea to put information like this which might be altered in a separate leaflet; it can then be updated without having to reprint and distribute the booklet for the whole scheme.]
2. Cars are replaced after three years or once they have completed 50,000 miles, whichever is sooner. Occasionally cars are returned before this time, eg because the holder has left the Company. In such cases, the car is reallocated until due for replacement; you may therefore be given a reallocated car rather than a new one.
3. If the delivery of a new car is delayed, a member of staff newly entitled to a car may be assigned a reallocated car (if one is available) until his or her new car is delivered. Members of staff who already have a company car should in these circumstances retain the old car until the new one is delivered.
4. Should you require optional extras beyond those supplied as standard with your choice of car, you should refer to leaflets CAR 1 and CAR 2 for the Company's policy on this.

2. Income Tax and your Company Car

1. Company cars are taxed not on the actual value of the benefit but according to a scale charge. Current levels of scale charge are shown in leaflet CAR 3. The scale charge is added to your taxable income and you are taxed on it at your marginal tax rate, as if you had received it as income.
2. As you will see, the scale charge increases:
    (a) according to engine size of the car
    (b) according to the market value of the car
    (c) if you use the car for less than a minimum number of business miles each year.

    If you wish your car benefit to be more tax effective, given a choice of engine size, you should choose a smaller engine size in a lower tax bracket.

3. The tax consequences of being provided with a company car are your responsibility and you must report this benefit when you complete your tax returns. The Company also has to report to the Inland Revenue on individuals who receive a company car.

3. Expenses the Company will Meet

1. As well as the cost of purchase and delivery of your company car, the Company will pay for:
   (a) its insurance
   (b) the annual road fund licence
   (c) routine servicing and repair
   (d) membership of a roadside breakdown organisation, eg the AA.
   When your company car becomes due for renewal, the Company will also arrange for its disposal.
2. The Company does not pay for petrol or oil. It does, however, reimburse you at a standard rate for the cost of these when you are using the car on company business. You will be notified of this rate from time to time.
3. You are responsible for any fines incurred by error of yourself or any other drivers of your company car. Fines should be settled promptly. Fines which have to be settled by the Company will be charged to you with £10 as administrative costs added.

4. Insurance

1. We operate a fleet insurance policy covering company cars. You may drive a company car only if you hold a full valid UK, EC or other driving licence which meets with the insurers' approval. The car is insured for you to drive or for other nominated drivers to drive with your permission. Under no circumstances must you or anyone known to you drive a company car if disqualified.
2. The car is insured for social, domestic and pleasure purposes and for use in connection with your employment with ABC. Any travel abroad should be notified to the personnel department who will provide insurance cover in the form of green cards. For approved business use, any costs associated with the provision of green cards will be paid for by the firm but private or holiday use costs will be borne by the user.

3. If you are involved in a car accident you must always inform the personnel department of the details, even if you do not expect an insurance claim to follow. If you wish to make an insurance claim, contact the personnel department who will advise you how to proceed.

[Employees should be provided with a Certificate of Motor Insurance and a statement of the important terms.]

5. Road Fund Licence

1. The personnel department will provide the road fund licence for your car and renew it when required; it will be sent to you on renewal to be affixed to your company car.
2. It is your responsibility to ensure that a current road fund licence is displayed in your company car.

6. Servicing and Repair

1. It is your responsibility to see that the car is properly serviced and maintained in a roadworthy condition.
    The Company has accounts with those dealers listed in the enclosed leaflet CAR 4 [not included in this example] and you should make arrangements for servicing and repair of your car with the appropriate dealer. All you are required to do is to sign the appropriate authorisation at the dealer to state that the work has been completed to a satisfactory standard and ABC will be invoiced direct.
2. You should note, however, that our contracts with these dealers do not permit them to effect major repairs or fit extras without prior authorisation from the personnel department. You should therefore notify the personnel department beforehand to seek clearance before booking your car in at a dealer for this work.

7. Disposal of Company Cars

1. When a company car becomes due for replacement, the personnel department will determine the price which might be obtained were the car disposed of to the motor trade (the "trade value").

Appendix

2. If you wish, you may buy your company car from the company at the trade value.
3. If you do not wish to buy it, the car will be offered to other members of staff, who will be invited to submit sealed bids for the car, with the highest bid receiving it. If no one wants the car or no bid exceeds the trade value, it will be sold to the trade.
4. Returned cars not yet due for replacement and which cannot be reallocated will also be offered to staff on the same basis as described in paragraph 7.3.

8. Accidents

1. At the Accident
   If anyone is injured, give first aid and call an ambulance. Do not move an injured person unless absolutely necessary.
   Do not admit liability to anyone. Answer questions by the police but sign a statement only if you want to. Send the personnel department a transcript of any statement you make.
   The following factors should be noted:
   (a) the measurements and conditions of the road
   (b) speed limits in operation and approximate speeds of vehicles involved, position of parked cars, road signs, weather, visibility.
   Take names and addresses and vehicle details of third parties and the names and addresses of any witnesses (a person who did not see the accident but who can comment on vehicle positions, etc can also be useful).
   Every accident must be reported to the personnel department, even if you do not expect an insurance claim to be made. Ignore suggestions that the accident should not be reported and refer any requests about the Company's insurance policy to the personnel department.
   If a driver is not capable of driving through injury or mental stress, the car should be moved to safety by a garage or another competent person.
2. Legal Position
   If damage or injury is caused to any person, animal, vehicle or other property, you must stop and if required to do so by any person having reasonable grounds, give your name and address and those of the owners of the vehicle. If for any reason you do not supply these details, you must report the accident to the Police as soon as possible

or in any case within 24 hours. Accidents involving personal injury must be reported to the personnel department at once. Never call upon or write to any injured person(s), their legal adviser, relations or friends. All communications received must be forwarded acknowledged to the personnel department.

CAR 1: Example of Restricted Car Entitlement

[This example applies to those who receive a company car because of job need or who are amongst the more junior managers who are eligible for a car on status basis.]

Car entitlement is as follows:

Grade	Type of car
7	Vauxhall Cavalier 1600 L
	Ford Sierra 1600 L
8	Vauxhall Cavalier 1600 GL
	Ford Sierra 1800 LX
9	Vauxhall Carlton 1.8 L
	Ford Sierra 2000i GLS
10	Vauxhall Carlton 2.0i GL
	Ford Granada 2.0 GL Auto

[The basis of this list is that there is an advance in model and style with grade. This serves to mark promotion. Cars in each category have been chosen with a similar cost new; they happen to have been chosen from just two of the fleet manufacturers. The advantage of a smaller range arises from economies of scale, eg discount because of bulk purchase, simplicity of maintenance, etc.]

The policy on optional extras is that all cars will be supplied with a car radio and rear seat belts where these are not fitted as standard. Cars in grades 9 and 10 may be supplied with a metallic finish if so desired. Other extras may be added but will be at the user's expense.

CAR 2: Example of Car Entitlement for Senior Managers

[This example applies to those more senior managers who receive a car primarily because of status.]

Car entitlement is on the basis of initial list price, as follows:

Grade 11–12	Up to £12,000
Grade 13–16	" " £15,000
Grades 17 and above	" " £18,000

The total cost of the vehicle (excluding delivery charges but including optional extras you select) should not exceed the price entitlement for your grade. You may choose any car which meets these price criteria, excluding:

(a) two seater cars
(b) secondhand cars
(c) convertible or soft top cars
(d) left hand drive cars.

You should notify the personnel department of your choice; the Company does, however, reserve the right to refuse to buy certain makes and models of car on the grounds that experience indicates that their depreciation and/or running costs are considerably more than other models. Please refer to the personnel department if you have any queries on this point.

Should you wish to have extras which would raise the initial cost of your car beyond your capital allowance these will be at your own cost. Some account will, however, be taken of their residual value when you relinquish the car.

[A suitable method might be to charge the user for the private use of the car an amount equal to the depreciation of the extras over the period of his or her use of the car. For example, Smith, whose capital entitlement is £11,000, wants extras worth an additional £720 on his car. It is agreed that these will be worth 50% of their new value when the car is sold. Smith is therefore charged £360 (50% × £720) over the period he has the car. If this was three years, it would be £10 per month. He would also have to pay VAT on this, ie a total cost of £11.75. This could be set against the scale rate tax charge, ie he would avoid tax otherwise payable on (12 × £11.75 =)£141 pa, ie he would save £56 tax pa (if his marginal tax rate was 40%); the actual cost of the extras would thus be only £85 pa to him.]

CAR 3: Current Income Tax Rates

[This leaflet would need to be updated to take account of any changes in the scale rates of car tax.]

Company cars are taxed not on the actual value of the benefit but according to a scale charge. Current levels of scale charge are shown below. The scale charge is added to your taxable income by your tax inspector and included in your tax assessment.

Scale Rates for Taxation of Company Cars 1992/93

Cars with an original market value of up to £19,250.

Cylinder capacity of car in cubic centimetres	Age of car at end of relevant year of assessment Under four years
1400 or less	£2,140
More than 1400 but not more than 2000	£2,770
More than 2000	£4,440

Cars with an original market value of more than £19,250.

Original market value of car	Age of car at end of relevant year of assessment Under four years
More than £19,250 but not more than £29,000	£5,750
More than £29,000	£9,300

These rates are modified according to the number of business miles that you travel in a tax year.

If you travel fewer than 2,500 miles on company business in your car in a year, the rates are increased by 50%.

If you travel more than 18,000 miles on company business in a year, the rates are halved.

INDEX

Note. The index covers Chapters 1 to 10 and the Appendix. Index entries are to page numbers. Index entries in bold type indicate figures or tables. Alphabetical arrangement is word-by-word, where a group of letters followed by a space is filed before the same group of letters followed by a letter, eg "tax treatment" comes before "taxable benefits". In determining alphabetical arrangement, initial articles and prepositions are ignored.

accidents
  control 48, 94
  in specimen car scheme booklet 97, 101–2
accounting treatment
  acquisition of cars 37–9
  contract purchase 33
acquisition of cars
  accounting treatment 37–9
  company tax treatment 39–42
  funding choices **28**
    buy now 29–30
    company policy 27–9
    use now – buy later 30–3
    use now – buy never 33–7
additional cars 62, 76
administrative costs 23
age of cars 58
agency arrangements 33
allocation *see* fleet allocation
annual budgets 21
auction of cars 51, 52–3, 55, 94–5

balance sheets 33, 38
balancing allowances *see* capital allowances
balancing charges *see* capital allowances
balloon leases *see* finance leases
bank borrowing 30–1

benefits-in-kind
  artificial scale 58
  car parking spaces 63
  chauffeurs 62
  definition 57
  employer loans to staff 89–90
  fuel for private use 69–70
  Government policy 2
  Inland Revenue approach 13
  personal contributions of staff 63–4
  sales of cars to staff 54–5
  in specimen car scheme booklet 98–9
  taxation of company cars
    present regime 57–64
    proposed new regime 64–9
borrowing, bank or finance house 30–1
budgets 4
  annual process 21
  benefits of budgetary control 21–2
  capital 22
  cash flow 22, 23–4
  comparisons 25
  creation 22–4
  forecasting 25
  long-range forecasts 24
  monitoring 24–5
  operating 22–3
  phasing 24

105

budgets (continued)
  revenue 22–3
  review 24–5
business cars 1, 2–3
  see also company cars
business mileage
  additional company cars 76
  business travel definition 74–5
  identification of 73–6
  mixed motive journeys 75–6
  national insurance calculations 76–9
  scale rate tax charges 59–60
buy now see acquisition of cars
buying out company cars 85–90

capital allowances
  car accessories 40
  contract purchase 32
  finance leases 41
  purchase of cars
    costing up to £12,000: 39
    costing over £12,000: 40
  reducing balance basis 39
capital budgets 22
capital gains tax 40
car allowances to staff
  background 86
  calculations
    adequacy of allowance 88–9
    base assumptions 86–7
    cost to company 87–8
    tax cost of car to employee 88
car fleets
  allocation see fleet allocation
  budgets see budgets
  costs 8, 23
  structure 8
car loan schemes
  employer loans 89–90
  external loans 90
car parking spaces 63
car specification 16, 45
cars, type and choice 8

cash allowances in lieu of company
  cars see car allowances to staff
cash flow budgets 22, 23–4
cash purchases 29–30
  company tax treatment 39–40
  VAT 80
chauffeurs 62
closed-end leases see finance leases
commercial aspects, company car
  policy 12
companies
  acquisitions of 9
  cars see company cars
  overheads see overhead costs
  policy needs 9
  tax treatment of, acquisition of cars
    39–42
company car policy
  choice of vehicles 45
  commercial aspects 12
  economic issues 10
  formulation 7–13
  ground rules 7
  measuring progress 7
  objectives
    achievement 10, 11–12
    setting 9–11
  operational aspects 12
  other issues 13
  political issues 10
  position statements 7
  present position, key questions 7–9
  reviews 7
  social issues 10–11
  strategic aspects 12
  tactical aspects 12
  technology issues 11
company car scheme specimen
  employee booklet 97–104
company cars
  buying out 85–90
  definitions 1, 57
  fleet audit see fleet audit

Index

company cars (continued)
  future 91–5
  industry 2–6
  introduction 1–6
  national insurance 76–9
  as "perks" 13, 16
  policy *see* company car policy
  and pool cars, differences between 61
  qualities required 45
  taxation *see* taxation of company cars
  VAT 80–3
Company cars – reform of income tax treatment (Government consultative document) 65, 91
contract hire 16, 34–5, 38
  against contract purchase 33
  corporation tax treatment 41–2
  VAT 81
contract purchase 32–3
  company tax treatment 39–40
  VAT 80–1
controls
  see also cost control
  budgetary 21–2
  fleet management 11
  waste due to lack of 9
corporation tax 32, 41–2
cost control
  data interpretation 43
  discounts 43
  insurance 47–9
  operating costs 45–6
  purchase price 44–5
  records 43
  replacement cycles 49–50, **50**
  residual value 47
  staff training 43
  standards 43
  whole-life vehicle cost 43, 44, **44**
costs
  see also cost control

budget creation 23–4
car fleet 8–9
maintenance *see* running and maintenance costs
running *see* running and maintenance costs
whole-life 8, 15, 43, 44, **44**
cylinder capacity
  proposed new tax regime 65
  scale rate tax charges 58–9

daily rental 33–4, 81
deferred purchase *see* hire purchase
depreciation 25, 30, 38, 45
  see also capital allowances
diesel cars
  fuel economy 45–6
  insurance 48
diesel fuel 91
disposal of cars
  auctions 51, 52–3, 55, 94–5
  capital allowances 40
  capital gains tax 40
  condition of cars 51
  fleet audit 94–5
  options 52
  part exchange 55
  pre-disposal preparation 55–6
  price factors 52
  sales to general public 55
  sales to staff 54–5
  in specimen car scheme booklet 100–1
  timing 51
  used car traders 53–4
  VAT 80

economic issues 10
employees *see* staff
employment levels for company cars 7–8
engine sizes *see* cylinder capacity

entitlement to company cars
  in specimen car scheme booklet 98, 102–3
environmental considerations 11, 46, 91–2
establishment costs 23
externally purchased services 9

finance, sources 9
finance house borrowing 30–1
finance leases
  balloon leases 36
  closed-end leases 36–7
  explanation 38
  fully-amortised leases 37
  open-end leases 37
  SSAP 21: 38
  VAT 81
Fixed Profit Car Scheme (FPCS) 72–3
fleet acquisition *see* acquisition of cars
fleet allocation
  advice to users 18–19
    specimen booklet 97–104
  contract hire 16
  cost 15–16
  grading personnel 15
  methods
    by grade 16
    by income 16–17
    parent company policy 16
    "user-chooser" 17
  as need to the business 16
  "perk" cars 16
  process of allocation 16–18
  selection and specification 18
  servicing 16
  specifying the car 16
fleet audit
  accident control 94
  fleet disposal 94–5
  fuel management 94
  objectives 93–4
  present status 94
  strategic economic issues 95
fleet disposal *see* disposal of cars
fleet funding
  choices 27–37
  strategy 8–9
fleet management
  benefits of getting it right 93
  company acquisitions 9
  control 11, 43
  in-house or external 24
  introduction 1–6
  specialist companies 43
    contract hire 35
    contract purchase 32, 33
  strategy 8
fleet managers 3–4
  accounting treatment, need for awareness of 37–8
  contract hire responsibilities 35
  expertise 43–4
  newly appointed 7
  product knowledge 11
FPCS (Fixed Profit Car Scheme) 72–3
fuel
  consumption 45–6
  diesel fuel 91
  management 94
  scale charges
    1992/93: **70**
    cost effectiveness 70, **71**
    private car fuel 69–70
    proposed new tax regime 67
  unleaded petrol 91
  VAT 81–3
fully-amortised leases *see* finance leases

goals, company car policy 9–12
Government
  revenue benefits 3
  scale rate tax charge increases 64
  tax neutrality policy 2, 65, 85

Index

grading
  cars 15–16, **17**
  personnel 15
  in specimen car scheme booklet 98

hire purchase 31
  company tax treatment 39–40
  VAT 80–1

ICTA *see* Income and Corporation Taxes Act
incentives
  driver award schemes 48
  Government
    for environmentally friendly choices 91–2
    for use of public transport 92
Income and Corporation Taxes Act 1988
  business travel 74–6
  chauffeurs 62
  company cars definition 57
  pool car conditions 61
income tax
  employees' liabilities *see* benefits-in-kind
  in specimen car scheme booklet 98, 103–4
  yield from company cars 3
Inland Revenue estimates of market size 1–2
insurance
  checklist 48
  instructions to drivers 47–8
  in specimen car scheme booklet 97, 99–100

K codes (PAYE) 68

lease purchase 31–2
  company tax treatment 39–40
  VAT 80–1
leasing *see* contract hire; daily rental; finance leases; operating leases
list prices *see* price bands in new tax regime
loan interest 40
loans
  for fleet purchase *see* borrowing, bank or finance house
  to staff *see* car loan schemes

maintenance costs *see* running and maintenance costs
market value 58, 60
mileage allowances
  company cars 70, 72
  private cars 72–3
  in specimen car scheme booklet 99
monthly reporting requirements 67–8
motor industry 2, 3
  links between companies 4–5, **6**

national insurance contributions
  collection 79
  company cars
    1992/93: **77–8**
    availability of car 79
    calculation of liability 78–9
    fuel for private use 69–70, **70**

objectives, company car policy 9–12
open-end leases 37
operating budgets 22–3
operating costs 45–6
operating leases 38
  see also contract hire
operational aspects, company car policy 12
option agreements 33
original market value
  definition 60
  scale rate tax charges 58
overdrafts *see* borrowing, bank or finance house
overhead costs 23, 43

own funds, use of *see* cash purchases

P11D *see* benefits-in-kind
part exchange on disposal 55
pay as you earn (PAYE) 68
penalties
   to drivers for insurance claims 48
   Government
      for environmentally unfriendly choices 92
   for use of cars instead of public transport 92
personal contributions *see* staff
personnel *see* staff
petrol *see* fuel
political issues 10
pool cars 61, 62
price bands in new tax regime 65–6
private mileage 73–6
purchase price 44–5
purchases *see* cash purchases; contract purchase; hire purchase; lease purchase

recruitment 10–11
reducing balance *see* capital allowances
replacement policies 49–50, **50**
residual value 25, **26**, 45, 47, 51
revenue budgets 22–3
roads, charges for use 92
running and maintenance costs
   escalation 50
   in specimen car scheme booklet 97
   variation 46
   VAT 81

sale of cars *see* disposal of cars
scale rate tax charges 64, 85
   1992/93: **59–60**
   additional cars 62
   basis of assessment
      age of car 58

cylinder capacity 58–9
original market value 58
business mileage 59–60
included items 60
national insurance calculations 78
in specimen car scheme booklet 98, 104
staff contributions 64
time apportionment 61
seniority *see* grading
servicing 16
   in specimen car scheme booklet 100
social issues 10–11
specification 16, 45
specimen employee booklet for company car scheme 97–104
SSAP 21: 32–3, 38
staff
   booklet for company car scheme, specimen 97–104
   company car needs 3
   costs 23
   loans *see* car loan schemes
   personal contributions
      national insurance liability 79
      points to consider 63–4
      VAT on contributions 64, 83
   sales of cars to 54–5
Statement of Standard Accounting Practice 21: 32–3, 38
strategic aspects, company car policy 12

tactical aspects, company car policy 12
takeovers 9
tax neutrality, Government policy 2, 65, 85
tax treatment, companies, acquisition of cars 39–42
taxable benefits *see* benefits-in-kind
taxation of company cars
   benefits-in-kind

Index

benefits-in-kind (continued)
  present regime 57–64
  proposed new regime 64–9
  scale rate tax charges 58–61
technology issues 11
timing
  cost benefits for tax purposes 22
  disposal of cars 51
  proposed new tax regime 68
toll roads 92
training
  fleet management staff 43
  drivers 48
tyres 46

unleaded petrol 91
use now – buy later *see* acquisition of cars
use now – buy never *see* acquisition of cars
used car traders 53–4

value added tax (VAT)
  car fuel 81–3
  cash or car schemes 83, 86
  contract hire 81
  contract purchase 32, 33, 80–1
  finance leases 81
  hire purchase 80–1
  lease purchase 80–1
  purchase 39, 80
  running costs 81
  sale 80
  staff contribution to running costs 64, 83
  temporary rental 81
vehicles *see* entries beginning with car

whole-life cost *see* costs
  windows of opportunity for vehicle replacement 49–50, **50**
writing-down allowances *see* capital allowances